Liquid Crystal Sensors

THE LIQUID CRYSTALS BOOK SERIES

Edited by

Virgil Percec
Department of Chemistry
University of Pennsylvania
Philadelphia, PA

The Liquid Crystals book series publishes authoritative accounts of all aspects of the field, ranging from the basic fundamentals to the forefront of research; from the physics of liquid crystals to their chemical and biological properties; and from their self-assembling structures to their applications in devices. The series will provide readers new to liquid crystals with a firm grounding in the subject, while experienced scientists and liquid crystallographers will find that the series is an indispensable resource.

PUBLISHED TITLES

Introduction to Liquid Crystals: Chemistry and Physics
By Peter J. Collings and Michael Hird

The Static and Dynamic Continuum Theory of Liquid Crystals: A Mathematical Introduction
By Iain W. Stewart

Crystals That Flow: Classic Papers from the History of Liquid Crystals
Compiled with translation and commentary by Timothy J. Sluckin, David A. Dunmur, and Horst Stegemeyer

Nematic and Cholesteric Liquid Crystals: Concepts and Physical Properties Illustrated by Experiments
By Patrick Oswald and Pawel Pieranski

Alignment Technologies and Applications of Liquid Crystal Devices
By Kohki Takatoh, Masaki Hasegawa, Mitsuhiro Koden, Nobuyuki Itoh, Ray Hasegawa, and Masanori Sakamoto

Adsorption Phenomena and Anchoring Energy in Nematic Liquid Crystals
By Giovanni Barbero and Luiz Roberto Evangelista

Chemistry of Discotic Liquid Crystals: From Monomers to Polymers
By Sandeep Kumar

Cross-Linked Liquid Crystalline Systems: From Rigid Polymer Networks to Elastomers
Edited By Dirk J. Broer, Gregory P. Crawford, and Slobodan Žumer

DNA Liquid-Crystalline Dispersions and Nanoconstructions
By Yuri M. Yevdokimov, V.I. Salyanov, S.V. Semenov, and S.G. Skuridin

Nanostructures and Nanoconstructions based on DNA
By Yuri M. Yevdokimov, V.I. Salyanov, and S.G. Skuridin

Bent-Shaped Liquid Crystals: Structures and Physical Properties
By Hideo Takezoe and Alexey Eremin

Liquid Crystal Sensors
By Albert Schenning, Gregory P. Crawford, and Dirk J. Broer

Liquid Crystal Sensors

Edited by
Albert P. H. J. Schenning
Gregory P. Crawford
and
Dirk J. Broer

CRC Press
Taylor & Francis Group
Boca Raton London New York

CRC Press is an imprint of the
Taylor & Francis Group, an **informa** business

CRC Press
Taylor & Francis Group
6000 Broken Sound Parkway NW, Suite 300
Boca Raton, FL 33487-2742

First issued in paperback 2022

© 2018 by Taylor & Francis Group, LLC
CRC Press is an imprint of Taylor & Francis Group, an Informa business

No claim to original U.S. Government works

ISBN-13: 978-1-498-72972-7 (hbk)
ISBN-13: 978-1-03-233951-1 (pbk)
DOI: 10.1201/9781315120539

Library of Congress Cataloging-in-Publication Data

Names: Schenning, Albert, 1966- | Crawford, Gregory Philip. | Broer, Dirk J.
Title: Liquid crystal sensors / [edited by] Albert Schenning, Gregory P. Crawford, and Dirk J. Broer.
Description: Boca Raton, FL : CRC Press, [2017] | Series: Liquid crystals book series | Includes bibliographical references and index.
Identifiers: LCCN 2017013419| ISBN 9781498729727 (hardback : alk. paper) | ISBN 9781315120539 (ebook)
Subjects: LCSH: Liquid crystal devices. | Liquid crystals. | Polymer liquid crystals. | Detectors.
Classification: LCC TS518 .L595 2017 | DDC 530.4/29--dc23
LC record available at https://lccn.loc.gov/2017013419

Visit the Taylor & Francis Web site at
http://www.taylorandfrancis.com

and the CRC Press Web site at
http://www.crcpress.com

Contents

Preface...ix
Editors...xi
Contributors xiii

Chapter 1   Bandwidth Tunable Cholesteric Liquid Crystal 1

                Jian Sun, Wanshu Zhang, Meng Wang, Lanying Zhang,
                and Huai Yang

Chapter 2   Photochromic Chiral Liquid Crystals for Light Sensing 33

                Ling Wang, Karla G. Gutierrez-Cuevas, and Quan Li

Chapter 3   Chiral Nematic Liquid Crystalline Sensors Containing
                Responsive Dopants 63

                Pascal Cachelin and Cees W. M. Bastiaansen

Chapter 4   Cholesteric Liquid Crystalline Polymer Networks as Optical
                Sensors 83

                Monali Moirangthem and Albert P. H. J. Schenning

Chapter 5   All-Electrical Liquid Crystal Sensors 103

                Juan Carlos Torres Zafra, Braulio García-Cámara, Carlos
                Marcos, Isabel Pérez Garcilópez, Virginia Urruchi, and José
                M. Sánchez-Pena

Chapter 6   Liquid Crystal-Integrated-Organic Field-Effect Transistors
                for Ultrasensitive Sensors 123

                Jooyeok Seo, Myeonghun Song, Hwajeong Kim, and
                Youngkyoo Kim

Chapter 7   Liquid Crystals in Microfluidic Devices for Sensing Applications..... 145

                Kun-Lin Yang

Index.. 159

Preface

This dynamic book is about the future. Its contributions advance our understanding of liquid crystals and their applications in many ways beyond well-known display technology. It is about the field's next stage and a possible revolutionary application—the liquid crystal sensor. Will this novel application of such soft matter materials eventually blossom into a technology as broad and global as the display? Liquid crystal sensors—lightweight, inexpensive, and often easy to fabricate—could become major players in the burgeoning industry of sensor technology.

No one could have predicted that the unusual phenomenon discovered by Friedrich Reinitzer nearly 130 years ago would launch a material employed around the world for its electro-optic behavior. Trying to measure the melting point of a cholesterol substance, the Austrian botanist and chemist observed two melting points—one characterized optically by a cloudiness followed by a transition into a more traditional transparent liquid. At first, he suspected impurities, but purification had no effect. He sought help from the German physicist Otto Lehmann, who proposed the intermediate state was a novel and unique kind of order and became the father of liquid crystal science.

Today, this intermediate state of matter, exhibited by thousands of substances, is well established and thoroughly studied. But perhaps even more impressive than its complex nature is the impact that liquid crystals have had worldwide. Reinitzer and Lehmann would be amazed to see how their discovery shaped the information age. Liquid crystal displays, in TVs several feet wide, hang on walls in family rooms, bedrooms, and hotel rooms; they sit on desktops in homes and offices; they shine from laptop computers. They generate some $100 billion a year.

Many alternatives have sought to displace liquid crystal displays, but the mature liquid crystal technology is ever advancing, with a growing research field and expanding market that it dominates for the foreseeable future. Now the question is: How can liquid crystals find application in other areas? What is the next liquid crystal revolution? Will it flourish as much as liquid crystal displays?

One area now gaining many liquid crystal scientists' attention is sensors. The innate fragility of the liquid crystal phase and its inherent sensitivity to various external stimuli offer intriguing possibilities for sensor applications. Small perturbations or disruptions locally can propagate and be amplified throughout the bulk material—a small disruption can result in a more macrosignal, such as an optical change.

One focus of contributions in this book is the cholesteric liquid crystal and its inherent pitch, bandwidth, handedness, sensitivity to external stimuli, and how small influences can trigger some large, often optical, measurable signals. These cholesteric materials, although more robust and stable today, are the same phase Reinitzer observed well over a century ago. The contributions also discuss liquid crystal polymer sensors, electro-optical-based sensors, and microfluidics.

With so many contributions from experts in the field, we are trying to discern whether sensors are the next technology thrust of liquid crystal materials.

Worldwide, sensors are a flourishing field with increasing applications, easily accessible by computers and the Internet. Mature sensor technologies are already competing for market share. For liquid crystals to break into this market in a significant way, they must demonstrate advantages over entrenched technologies. Some advantages revealed in these contributions include solution processability, including the potential to print them precisely with an inkjet printer, polymer encapsulation, spray coating, and simple spin coating techniques that process thin layers on substrates. Many of these attributes and processes may open a way into the established sensor market for liquid crystals because they offer easy and inexpensive processing, without using large-scale production and expensive processing equipment, and can be integrated into standard manufacturing.

It took nearly 80 years for the liquid crystal phase to be transformed into something really practical, starting with simple displays on wristwatches and calculators, but only a few decades to turn into a game-changing technology that enabled portable and desktop computing and large-area entertainment displays. Will liquid crystals be the next revolution in sensor technology? It is still early, but we believe there is real potential for liquid crystal sensors someday to have that same global impact that displays now provide.

We are confident you will enjoy the following contributions as much as we have.

Albert P. H. J. Schenning
Gregory P. Crawford
Dirk J. Broer

Editors

Albert P. H. J. Schenning is a professor at the Eindhoven University of Technology. His research interests center on functional organic materials. Schenning earned his PhD at the University of Nijmegen in 1996 on supramolecular architectures based on porphyrin and receptor molecules with Dr. M.C. Feiters and Professor Dr. R.J.M. Nolte. Between June and December 1996, he was a postdoctoral fellow in the group of Professor Dr. E.W. Meijer at Eindhoven, University of Technology working on dendrimers. In 1997, he joined the group of Professor Dr. F. Diederich at the ETH in Zurich, where he investigated conjugated oligomers and polymers based on triacetylenes. From 1998 until 2002, he was a Royal Netherlands Academy of Science (KNAW) fellow at Eindhoven University of Technology (Laboratory of Macromolecular and Organic Chemistry) active in the field of supramolecular organization of pi-conjugated polymers. In 2010 he joined the group of Professor Dr. H. Coles and Dr. S. Morris at the University of Cambridge, for a sabbatical stay, working on cholesteric liquid crystals. He received the European Young Investigators Award from the European Heads of Research Councils and the European Science Foundation, the golden medal of the Royal Dutch Chemical Society and a Vici grant from the Netherlands Organisation for Scientific Research (NWO). In total, he has around 300 publications in peer-reviewed journals and 5 patents.

Gregory P. Crawford is professor of physics and president of Miami University. Prior to Miami University, he was dean of the College of Science and vice president and associate provost at the University of Notre Dame, and dean of Engineering at Brown University. In addition to his academic career which began in 1996, Crawford has co-founded two biotechnology start-up companies.

Crawford, a physicist, earned his bachelor's (mathematics and physics), master's (physics), and doctorate (chemical physics) degrees from Kent State University. As a graduate student, he completed his PhD at the Liquid Crystal Institute. Crawford completed two postdoctoral fellowships and was a researcher at the Xerox Palo Alto Research Center before joining the faculty at Brown University. He applied his research background in optics, photonics, and soft matter materials to address areas related to human health, for example, the analysis and dating of bruises, which are often key pieces of evidence in child abuse cases. Crawford's research career began in the flat panel display field and understanding the fundamental role of liquid crystal–substrate interactions. His work includes more than 400 research and education publications, review articles, and book chapters, and 21 US patents and patent applications. In addition to his research interests, Crawford dedicated his academic career to entrepreneurial education paradigms—bringing science, technology, and business together for students to create value out of basic university research.

Dirk J. Broer is a polymer chemist (PhD, University of Groningen) and specialized in polymer structuring and self-organization. He worked at Philips Research in Eindhoven, The Netherlands from 1973 to 2010. The typical research topics were vapor phase polymerization, optical data storage, telecommunication fibers, and liquid crystal networks. He started his work on liquid crystal materials in 1985 developing the process of in-situ photopolymerization of liquid crystal monomers to form densely crosslinked and monolithically ordered liquid crystal networks. At Philips Research from 1991 he developed optical films and components for LCD enhancement based on this technology. Most of the flat panel liquid display screens make use of the materials he developed. In 2000 he started his work on new manufacturing technologies of LCDs for large area displays and electronic wallpaper. He invented the paintable display technology. From 2003 to 2010, Dirk J. Broer was appointed as senior research fellow and vice president at the Philips Research Laboratories specializing on biomedical devices and applications of polymeric materials.

From 1996 Dirk J. Broer is a professor at the Eindhoven University covering research topics as liquid crystal orientation, polymer waveguides, solar energy, organic semiconductors, nanolithography, soft lithography, and polymer actuators for biomedical microfluidic systems. From 2010 he is a fulltime professor in Eindhoven specializing in functional organic materials for clean technologies as energy harvesting, water treatment, and healthcare applications. Special emphasis is given to liquid crystal responsive polymers that morph under the action of heat, light, electrical- and magnetic fields or driven by contacts with agents and/or changes in their ambient conditions and to membrane technology with monodisperse nano-porosity.

Dirk J. Broer is a member of Royal Dutch Academy of Sciences. In total, he has around 260 publications in peer-reviewed journals and more than 120 US patents.

Contributors

Cees W. M. Bastiaansen
Faculteit Scheikundige Technologie
Technische Universiteit Eindhoven
Eindhoven, The Netherlands

Pascal Cachelin
School of Engineering and Materials
Science
Queen Mary University of London
London, United Kingdom

Juan Carlos Torres Zafra
Displays and Photonics Applications
Group (GDAF-UC3M)
Electronic Technology Department
Carlos III University of Madrid
Leganés, Madrid, Spain

Braulio García-Cámara
Displays and Photonics Applications
Group (GDAF-UC3M)
Electronic Technology Department
Carlos III University of Madrid
Leganés, Madrid, Spain

Karla G. Gutierrez-Cuevas
Liquid Crystal Institute and Chemical
Physics Interdisciplinary Program
Kent State University
Kent, Ohio

Hwajeong Kim
Organic Nanoelectronics Laboratory
KNU Institute for Nanophotonics
Applications (KINPA)
Department of Chemical Engineering
School of Applied Chemical Engineering
and
Priority Research Center
Research Institute of Advanced Energy
Technology
Kyungpook National University
Daegu, Republic of Korea

Youngkyoo Kim
Organic Nanoelectronics Laboratory
KNU Institute for Nanophotonics
Applications (KINPA)
Department of Chemical Engineering
School of Applied Chemical
Engineering
Kyungpook National University
Daegu, Republic of Korea

Quan Li
Liquid Crystal Institute and Chemical
Physics Interdisciplinary Program
Kent State University
Kent, Ohio

Carlos Marcos
Thyssenkrupp Elevator
Manufacturing
Móstoles, Madrid, Spain

Monali Moirangthem
Functional Organic Materials and
Devices
Eindhoven University of
Technology
Eindhoven, The Netherlands

Isabel Pérez Garcilópez
Displays and Photonics Applications
Group (GDAF-UC3M)
Electronic Technology Department
Carlos III University of Madrid
Leganés, Madrid, Spain

José M. Sánchez-Pena
Displays and Photonics Applications
Group (GDAF-UC3M)
Electronic Technology
Department
Carlos III University of Madrid
Leganés, Madrid, Spain

Jooyeok Seo
Organic Nanoelectronics Laboratory
KNU Institute for Nanophotonics
    Applications (KINPA)
Department of Chemical Engineering
School of Applied Chemical Engineering
Kyungpook National University
Daegu, Republic of Korea

Myeonghun Song
Organic Nanoelectronics Laboratory
KNU Institute for Nanophotonics
    Applications (KINPA)
Department of Chemical Engineering
School of Applied Chemical Engineering
Kyungpook National University
Daegu, Republic of Korea

Jian Sun
Department of Materials Science
    and Engineering
University of Science and Technology
    Beijing
Beijing, China

Virginia Urruchi
Displays and Photonics Applications
    Group (GDAF-UC3M)
Electronic Technology Department
Carlos III University of Madrid
Leganés, Madrid, Spain

Ling Wang
Liquid Crystal Institute and Chemical
    Physics Interdisciplinary Program
Kent State University
Kent, Ohio

Meng Wang
Department of Materials Science
    and Engineering
College of Engineering
Peking University
Beijing, China

Huai Yang
Department of Materials Science
    and Engineering
College of Engineering
Peking University
Beijing, China

Kun-Lin Yang
Department of Chemical and
    Biomolecular Engineering
National University of Singapore
Singapore

Lanying Zhang
Department of Materials Science
    and Engineering
College of Engineering
Peking University
Beijing, China

Wanshu Zhang
Department of Materials Science
    and Engineering
University of Science and Technology
    Beijing
Beijing, China

1 Bandwidth Tunable Cholesteric Liquid Crystal

Jian Sun, Wanshu Zhang, Meng Wang, Lanying Zhang, and Huai Yang

CONTENTS

1.1 Introduction 1
1.2 Broadening the Reflection Bandwidth...3
1.3 Stimuli-Responsive Reflection Band 10
1.4 Surpassing the Reflectance Limitation... 18
1.5 Conclusion and Outlook 23
References...25

1.1 INTRODUCTION

The cholesteric phase of liquid crystals (LCs) presents a macroscopic helical structure, which was first observed by Reinitzer in 1888.[1] Since then, the cholesteric phase, also known as the chiral nematic phase, can be introduced by external chiral molecules in the nematic phase rather than cholesterol. As shown in Figure 1.1, the spatial orientation of the rod-shaped molecules (or mesogens) spontaneously rotates by a constant angle along the direction of helical axis in cholesteric liquid crystals (CLCs). The helical sense of the director is nonsuperimposable on its mirror image; thus, it is determined by the configuration of the chiral group within the molecule and it can be expressed as the positive sign (+) for right-handed (RH) helix and the negative sign (−) for left-handed (LH) helix, respectively. The helical pitch, p, is the distance over which the LC molecules rotate 2π twist. This periodic helical arrangement of the molecules causes unique selective reflection of light according to the Bragg regime.

The helical "structural colors" of CLC, distinguishing from the pigment, is omnipresent in living matter.[2-5] The Bragg reflection wavelength λ_0 is defined by $\lambda_0 = n \times p$, where $n = (n_o + n_e)/2$ is average refractive index of LCs, while n_o and n_e are the ordinary and extraordinary refractive indices, respectively. The reflection bandwidth $\Delta\lambda$ is given by $\Delta\lambda = \lambda_{max} - \lambda_{min} = (n_e - n_o) \times p = \Delta n \times p$, where Δn is the birefringence of the LC molecules. When unpolarized light passes through CLC, the circularly polarized light of same handedness gets reflected, whereas opposite handedness gets transmitted, leading to only 50% reflection of unpolarized light (Figure 1.2). Moreover, the wavelengths of reflected light depend on pitch of CLC. Therefore, the helical pitch, p, and the helical orientation are the two important parameters to determine the reflection of light.

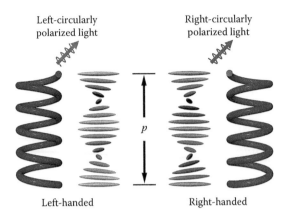

Left-circularly polarized light

Right-circularly polarized light

p

Left-handed

Right-handed

FIGURE 1.1 Schematic representation of helical structure of the cholesteric liquid crystal phase with different handedness.

CLCs can be obtained in two different ways. First, by having the molecules consisting of the cholesteric phase, intrinsically. And the second way is to dissolve chiral molecules into achiral nematic LC host, and the nematic phase is transformed into a cholesteric phase. The ability of a chiral dopant to induce twist deformation of an achiral nematic LC is defined as helical twisting power (HTP, value β). This inherent parameter of chiral dopant depends on the host–guest combination in LC medium,[6] and is quantified as: $\beta = 1/(p \times e.e. \times c)$, where $e.e.$ is the enantiomeric excess of chiral dopant and c is the chiral dopant concentration. To obtain a certain pitch length or reflective wavelength of CLC, it is possible to adjust the doping concentration of the chiral molecule by the above equation. The advantage of the second method is that the properties of CLCs can be tuned easily depending on the requirement of the particular application.

In general, $\Delta\lambda$ displays the width of the bandgap at half height within 100 nm because Δn is typically less than 0.3. This narrow optical bandgap together with reflection of only one polarization of light makes a major limitation in many areas

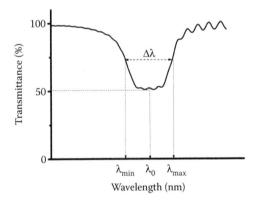

FIGURE 1.2 The selective light reflection property of the cholesteric liquid crystal phase.

of application, such as reflective colored displays, optical communication devices, and energy-saving windows. Herein, this chapter focuses on the cholesteric liquid crystalline materials presenting the enhancement in the reflection bandwidth across infrared (IR), visible, and ultraviolet (UV) region, which is fabricated in the presence of external stimuli such as mechanical pressure, temperature, electric/magnetic field, and light irradiation. This enhancement in the reflection bandwidth could be very useful for the applications of fiber-optic communications, laser emissions, integrated optics, biosensing, photo-activation, and laser protection.[7–9]

1.2 BROADENING THE REFLECTION BANDWIDTH

As mentioned above, the reflection bandwidth $\Delta\lambda$ of a uniformly pitched CLC is usually limited to few tens of nanometers due to limited Δn. Narrow reflection bandwidth is the limitations for many of the applications such as full-color reflective polarizer-free displays, broadband polarizers, light enhancement films, smart switchable windows, or IR-stealth, the bandwidth must be dramatically increased. Therefore, extensive investigations have been made on the wide-band reflection of CLCs in the decades.

To significantly increase the bandwidth of the selective reflection, a helical pitch that varies along the director is required. Then, the bandwidth is mainly determined by the pitch difference Δp, that is, $\Delta\lambda = n \times \Delta p$. Thus, CLC films having nonuniform pitch distribution (gradient or nongradient pitch distribution) can substantially broaden the reflection bandwidth. Based on the idea, novel CLC materials and innovative cholesteric-liquid-crystalline architectures have been developed.

As it is known, the LC structures can be frozen by photopolymerization of LC monomers.[10–15] By the selection of monomers and the optimization of curing conditions, desired mesogenic order can be frozen by the polymer network formed upon UV irradiation. When the UV-light intensity is locally controlled throughout a sample, composition gradients in the blends may occur due to the diffusion of reactive mesogens, creating local changes in properties such as refractive index, birefringence, or switching ability. Thus spatial modulation of structures and complex molecular arrangements of LC can be achieved. Photoinduced diffusion with simultaneous polymerization in a layer of CLC materials has been an elegant tool for creating pitch gradient.

Broer et al. firstly reported the method to achieve a CLC film with broad bandwidth over the entire visible spectrum.[16–25] The materials used in their research comprise CLC diacrylate monomer, nematic liquid crystal (NLC) monoacrylate monomer, photoinitiator, and UV-absorbing dye, as shown in Figure 1.3a.[22] Due to the presence of a UV-absorbing dye, a UV-intensity gradient occurred in the transverse direction. Therefore, the polymerization at the surface close to the source is much faster than at the bottom of the layer. Since the UV reactivity of the CLC diacrylate monomer is higher than that of NLC monoacrylate monomer, depletion of the diacrylate near the top generates a concentration gradient, causing diffusion of diacrylate toward the side of the composite layer with stronger UV intensity (Figure 1.3b).[25] To this material system, the pitch length decreases with increasing fraction of a chiral component

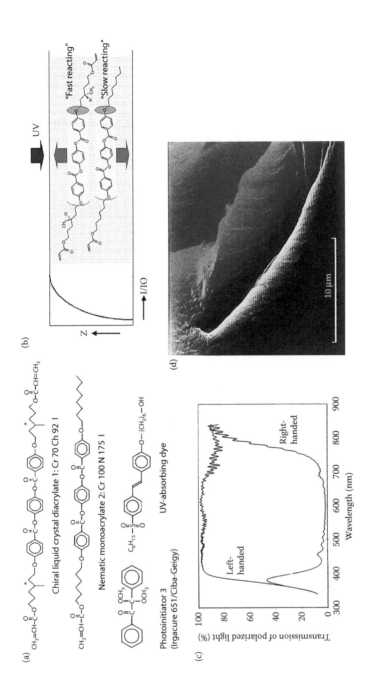

FIGURE 1.3 (a) Cholesteric diacrylate, nematic monoacrylate, photoinitiator, and UV-absorbing dye used in Reference 22 to fabricate a cholesteric elastomer with a pitch gradient. (b) Representation of the gradient of UV intensity in the mixture (left) and direction of diffusion of different monomers during the curing. (Reproduced with permission from Broer, D. J.; Lub, J.; Mol, G. N. *Mol. Cryst. Liq. Cryst.* 2005, 429, 77. Copyright 2005, Taylor & Francis Inc.) (c) Transmission of circularly polarized light of a 15-μm-thick film of a cholesteric network with pitch gradient. (d) Scanning electron microscopy image of the fracture surface of a pitch-gradient cholesteric network. The bands increase linearly in thickness from the bottom to the top. (Reprinted by permission from Macmillan Publishers Ltd. *Nature*, Broer, D. J. et al., 378, 467, copyright 1995.)

in the CLC composite, causing the pitch to vary over the cross section; thus, the prepared polymer film has a pitch gradient of about 270–450 nm, reflecting incident light with the wavelength range of 400–750 nm, as shown in Figure 1.3c and d.[22] This method to fabricate a pitch gradient increasing from top to bottom in the CLC film by controlling the kinetics of photopolymerization has been commonly used to broaden the reflection bandwidth of CLC.

In order to create a UV-light-intensity gradient over the film thickness, a UV-absorbing dye is usually required. Mitov et al.[26,27] discovered that the LC constituent of the gel possessed natural absorbing properties; thus, a UV-intensity gradient can occur in the transverse direction without the intervention of dye. The broadband reflective liquid crystalline gels can be obtained under asymmetrical conditions for the irradiation. The materials used in the research comprise a room-temperature CLC, bifunctional photoreactive nematic mesogens, and photoinitiator. Before curing, $\Delta\lambda$ for the composite is about 80 nm. After curing by irradiation of a UV-light beam on one side of the cell, $\Delta\lambda$ is significantly increased to about 220 nm, as shown in Figure 1.4a. When the cell is irradiated by UV light on both sides at the same time, the phenomenon of broadening of the reflection band was much less significant compared to the asymmetrical illumination conditions. Transmission electron microscopy (TEM) images of the cross sections indicate a network of oriented polymer chains and a polymer gradient from the top to the bottom of the cell, especially when the irradiation conditions are asymmetrical, as shown in Figure 1.4b.[26] Moreover, a symmetric or nonsymmetrical broadening can be obtained around the central wavelength by controlling curing conditions. This method is also applied to different CLC blends, and generic nature of the procedure was demonstrated.[28–30]

The variation of chiral dopants to induce the pitch gradient throughout the cross section of the film can also be achieved in other ways. For example, Mitov et al. firstly fabricated a pitch gradient in a cholesteric (Ch) structure using glass-forming Ch oligomers by thermal diffusion between CLC layers with different pitch length.[31–37] The structure of molecule used is shown in Figure 1.5a, and the reflected light ranging from blue to red can be simply tuned by changing the molar percentage of chiral mesogens in the oligomer molecule. The fabrication procedure of the films with pitch gradient is schematically illustrated as shown in Figure 1.5b. Two glass plates that were separately coated with thin Ch films reflect red ($\lambda_1 = 710$ nm) and blue colors ($\lambda_2 = 445$ nm), respectively, and were used to fabricate the cell. By controlling thermal diffusion of molecules in the cell at the temperature above the glass transition temperature T_g, a transverse concentration gradient was generated and subsequently pitch gradient was established. After quenching it to room temperature, the pitch gradient can be frozen in the glassy solid film. Moreover, the bandgap features can be tuned simply by varying the annealing time. By this simple method, the wavelength bandwidth can be broadened by more than 300 nm with the transmittance that exhibits a broad bandgap with a plateau inthe visible spectrum, as shown in Figure 1.5c. The TEM cross section of film shows that a regular and continuous transverse pitch gradient, which means the periodicity, smoothly changes from the red to the blue part of the film (Figure 1.5d).

Furthermore, due to the limitations that the helical structure of a conventional glassy CLC is strongly influenced by fluctuations of the temperature and is stable only at temperatures much lower than the glass-transition temperature, an alternative

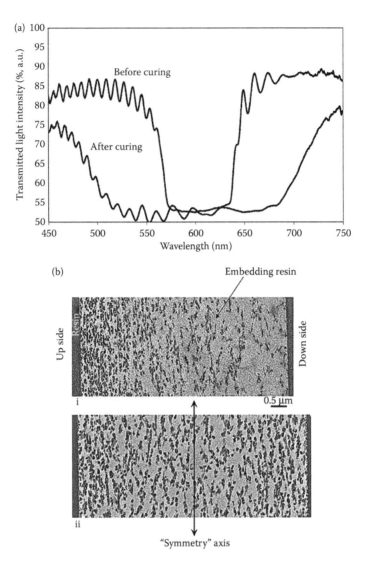

FIGURE 1.4 (a) Transmittance of a polymer-stabilized cholesteric liquid crystal before and after curing when the cell is irradiated with UV light from a single side. (b) TEM cross sections of polymer network of polymer-stabilized cholesteric liquid crystals after removal of the LC and embedding in a resin matrix. The cell was irradiated under: (i) from a single side and (ii) from both sides simultaneously. (Reprinted with permission from Relaix, S.; Bourgerette, C.; Mitov, M. *Appl. Phys. Lett.* 2006, 89, 291507. Copyright 2006, American Institute of Physics.)

to this method consists of carrying out a polymerization of pitch gradient, thus the pitch gradient of broad reflection bandwidth is formed and kept effectively and permanently, improving the thermal stability largely.[38–40]

On the other hand, the pitch gradient can also be induced by concentration diffusion of chiral dopants between layers. Sixou et al. obtained a CLC film with broad

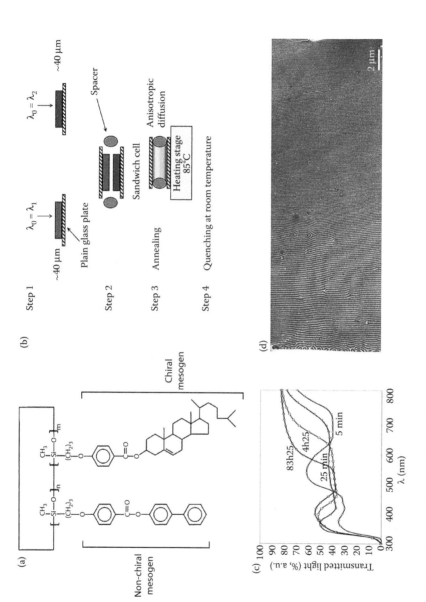

FIGURE 1.5 (a) General formula of cholesteric polysiloxane oligomer used in Reference 31. (b) Experimental procedure leading to a vitrified CLC structure with a pitch gradient from two single-pitch CLC films. *(Continued)*

FIGURE 1.5 (Continued)   Cross sections are represented. Step 1: Two cholesteric films with red ($\lambda_1 = 710$ nm) and blue colors ($\lambda_2 = 445$ nm) are coated on separate glass plates. Step 2: A sandwich cell is fabricated with these open films. Step 3: The cell is annealed for a variable time above the glass-transition temperature. A transverse concentration gradient occurs. Step 4: The cell is finally quenched at room temperature. The concentration gradient, and consequently the pitch gradient, is frozen inside a solid film. (c) Transmittance of bilayer cholesteric film at different annealing times during Step 3 of Figure 1.5b. The bandgap is tunable by controlling only the annealing time. The change is irreversible. (Mitov, M.: *Adv. Mater.* 6260. 2012. Copyright Wiley-VCH Verlag GmbH & Co. KGaA. Reproduced with permission.) (d) TEM cross-section of the cholesteric film after 25 min annealing and quenching at room temperature. The fingerprint texture exhibits a continuous periodicity gradient. (Reprinted with permission from Zografopoulos, D. C. et al. *Phys. Rev. E* 2006, 73, 061701. Copyright 2006 by the American Physical Society.)

bandwidth in the cell assembled by two glasses that were coated with nematic or cholesteric monomer films. The composition gradient in the cell induces diffusion over the film thickness leading to a gradual change of the helical pitch over the cross section of the cell. The texture containing the pitch gradient is then blocked by photopolymerization. In the method, the factors of HTP of chiral dopant, concentration, and contact time play an important role.[41,42]

All the above methods to broaden reflection bandwidth of CLC are based on creating pitch-gradient distribution. Distinguishingly, Yang et al. reported a CLC gel film with broad bandwidth by creating a nonuniform pitch distribution through mixing polymerizable particles with different pitches.[43–46] This method does not require complicated synthesis of the CLC diacrylate monomer and precise control of UV-intensity gradient, while the bandwidth and the location of the reflection band can be controlled accurately according to experimental design. For example, the materials used are photopolymerizable monomer (C6M)/chiral dopant (ZLI-4572) composites that mixed uniformly by solvent evaporation. The desired mixture powders were prepared through crush and filtration in crystalline (Cr) state at room temperature. Adjusting the concentration of ZLI-4572, the particles with different pitch length can be obtained, and they exhibit phase transition from Cr to Ch at about 355 K. By mixing particles with different pitches in the Cr phase in different weight proportion and cross-linking the LC monomer molecules by photopolymerization in the planar oriented Ch phase, CLC films with nonuniform pitch distributions are obtained. Moreover, the bandwidth of the reflection spectrum and the location of reflection band of the composite films can be controlled accurately by controlling the pitch lengths of the Ch phase of the particles, as shown in Figure 1.6.[43,44]

Based on the methodology of Broer, a novel method to fabricate asymmetrical superwide pitch-gradient distribution is proposed by controlling mesogenic phase structural transition of photopolymerizable acrylate monomer mixtures that presented phase transition between Ch and smectic (SmA) phases.[47] It has been known that in a small temperature interval just above the transition from Ch to SmA, a drastic change in pitch may occur, which is the result of the formation of SmA-like short-range ordering (SSO) structure.[48–51] The materials used in the research are shown

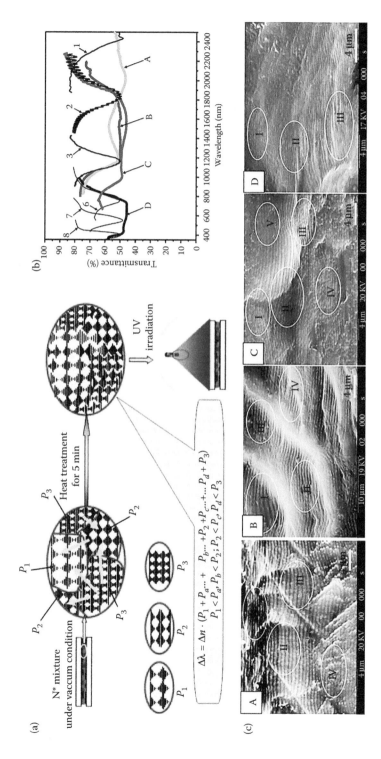

FIGURE 1.6 (a) Schematic representation of the preparation of an N* LC composite film with nonuniform pitch distribution. (Reproduced with permission from Huang, W. et al. *Liq. Cryst.* 2008, 35, 1313. Copyright 2008, Taylor & Francis Inc.) (b) Transmittances of samples of particles 1–3 and 6–8 and cross-linked mixture A–D as a function of wavelength. (Reprinted with permission from Bian, Z. et al. *Appl. Phys. Lett.* 2007, 91, 201908. Copyright 2007, American Institute of Physics.) (c) The microstructures of the freeze-fractured surface of the CLC composite films prepared from mixtures A–D. (Reproduced with permission from Huang, W. et al. *Liq. Cryst.* 2008, 35, 1313. Copyright 2008, Taylor & Francis Inc.)

in Figure 1.7a, and the fabricating procedure of the films is illustrated in Figure 1.7b, where the nematic monoacrylate, nematic diacrylate, smectic monoacrylate are denoted as NM, ND and SM, respectively. At the initial state, the cell is at the temperature of CLC phase to adopt a planar orientation. Upon UV-light irradiation, a light-intensity gradient and the upward diffusion making a concentration gradient of ND, as illustrated in Broer's reports. Differently, ND concentration gradient causes a gradient of the SSO. Then a nanostructure transition from Ch to SSO may occur at the film depth at which the transition temperature and irradiation temperature meet ($T_{SSO} = T_I$). Moreover, a decrease in ND concentration (curve *a*) results in a relative increase of SM concentration, thus a decrease of pitch length in Ch layer. For SSO, the pitch length increases rapidly with decreasing temperature, leading to the formation of SSO layer with a large pitch length gradient. By controlling the curing temperature to be close to SmA–Ch phase transition, a novel architecture that combined Ch and SSO nanostructures can be obtained. To achieve this kind of films, the formation of SSO is a crucial factor. As shown in Figure 1.7c and d, the bandwidth of Film 0 that without UV-absorbing dye is only about 170 nm and subtle pitch gradient is observed from scanning electron microscopy (SEM) images. Film 1a that irradiated at a temperature close to SSO phase has a superwide pitch length distribution. It exhibits an extremely broad bandwidth of almost 11 µm. While Film 1b which is fabricated by curing at 15 K above the Ch–SmA phase transition has a much smaller pitch gradient than that of Film 1a. Furthermore, by adjusting the concentration of monomers, the transmission spectrum of a polymeric film can be changed. For example, Film 2 (Figure 1.7c) possesses an ultra-broad bandwidth of about 13 µm, and it is the widest reflection bandwidth that has been reported so far.

1.3   STIMULI-RESPONSIVE REFLECTION BAND

CLCs are smart materials with stimuli-response function. Their helical pitch, *p*, can be precisely modulated by endowing a variety of external and dynamic stimuli such as electric field,[52] heat,[53] and light.[54] The controllability of optical performances, including reflection wavelength shifting, reflection bandwidth change, and nonreflection mode switching, enable CLCs as an important class of intelligent materials.

Among various external stimuli, the electrically controllable CLCs are the most promising for real technological applications. Till date, the electrically controllable dynamic optical responses have been examined in both nonpolymerizable and polymer-stabilized CLCs (PSCLCs) to enable electric field switching or pitch distortion for reflection bandgap switching,[55,56] tuning,[57] and broadening.[58–61]

Lavrentovich et al.[62] proposed an oblique helicoid structure of CLCs to produce electrically tunable selective reflection of light in an exceptionally wide spectral range from UV to near-IR (360–1520 nm). The applied electric field acts along the helical axis and tends to realign the LC molecules along itself. Upon adjustment of the amplitudes of the electric field, the pitch *p* is changed, thus leading to a precise modulation of the wavelength and bandwidth of the selective reflection peak (Figure 1.8a). As shown in Figure 1.8b, with reducing the electric field, the CLCs show a sequence of redshift in the wavelength of reflection band, from UV to visible

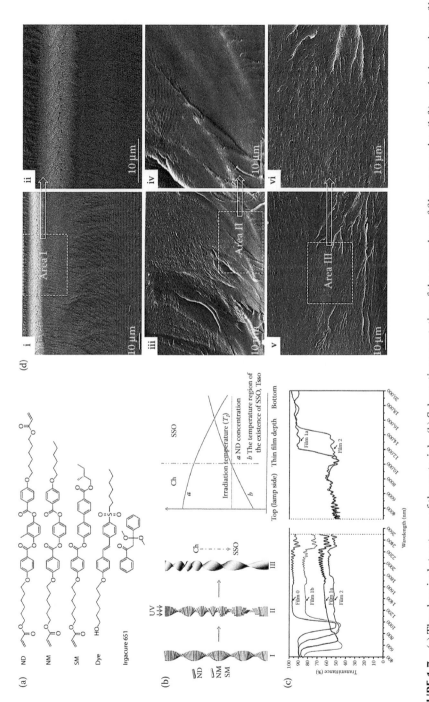

FIGURE 1.7 (a) The chemical structures of the materials. (b) Schematic presentation of the procedure of film preparation (left), and schematic profiles of the ND concentration *(Continued)*

FIGURE 1.7 (Continued) *a*, the temperature region of the existence of the SSO *b* and irradiation temperature T_I versus the thin-film depth (right). The film having Ch and SSO is formed when T_I is just above the temperature region of the existence of SSO, T_{SSO}. (c) Transmission spectra of Films 0, 1a, 1b, and 2. (d) SEM images of the fractured surface of Films 1a, 1b, and 0. (i) SEM image of Film 1a shows a very great pitch gradient; (ii) magnified SEM image of Area I; (iii) SEM image of Film 1b; (iv) magnified SEM image of Area II also shows a pitch gradient; (v) SEM image of Film 0; (vi) magnified SEM image of Area III shows subtle pitch gradient. (Reproduced with permission from Zhang, L. et al. *Liq. Cryst.* 2016, 43, 750. Copyright 2016, Taylor & Francis Inc.)

blue, then green, orange, red, and finally, near IR. This was also further confirmed by polarizing optical microscope textures shown in Figure 1.8c.

A fast electrically switchable reflectors with narrow and broad bandwidth are developed by Hikmet et al.[63] in cross-linked-network-dispersed cholesteric gels, which is done by *in situ* polymerization of a LC monoacrylate and diacrylate mixture in the presence of nonreactive LC molecules. The network can provide a memory state of the initial cholesteric configuration upon application of an electric field. In this process, micro-phase separation leads to concentration fluctuations corresponding to regions with different pitches. Figure 1.9a shows the switching modes of reversible change in the reflection characteristics obtained for cholesteric gels. The reflection band shifts gradually to low wavelengths on increasing voltage. On further increasing the voltage, the reflectance starts decreasing with complete disappearance at 60 V. This behavior is associated with the tilting of the cholesteric helix followed by the unwinding of itself. Furthermore, as shown in Figure 1.9b, introduction of excited-state quenchers, which reduce radical formation and cause an inhomogeneous intensity of light distribution, can induce an inhomogeneous distribution of pitch; hence, results in formation of a broader reflection band. These broadband cholesteric gels can be switched reversibly between silver-colored (reflecting) and transparent (nonreflecting) states (Figure 1.9c). Moreover, based on this mechanism, Schenning et al. develop an electrically switchable broadband IR reflector via photoinduced diffusion during photopolymerization. The polymerized network is of crucial importance because it simultaneously creates the broadband IR reflection and stabilizes the planar orientation of the nonreactive LC component. As a result, the fabricated broadband IR reflectors can be switched manually between IR reflective and transmission modes on applying an electric voltage to respond to changes in environmental conditions, which have a great potential as smart window to control interior temperatures and save energy.

Yang et al.[64] put forward an innovative electrically switched CLC composite by doping chiral ionic liquid (CIL). Upon an electric field applied, the anions and the cations of CIL moved toward the anode and the cathode, respectively, thus forming a density gradient of the chiral groups throughout the thickness of the cell, which resulted in formation of the broad reflection band. As shown in Figure 1.10a and b, this composite can be switched electrically between the transparent, scattering, and reflecting state in the visible region. All these states can be memorized for more than 7 days after the applied electric field has been turned off. Furthermore, the bandwidth of the reflection can be controlled accurately and reversibly by adjusting the intensity of DC electric field followed by AC electric field (Figure 1.10c). One of the potential applications of this composite is E-paper which exhibits a brilliant color by

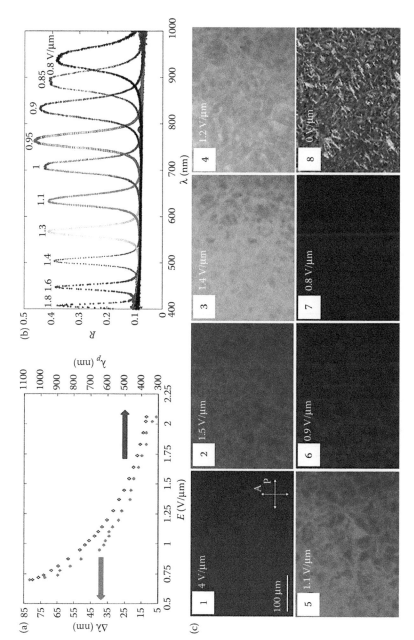

FIGURE 1.8 **(See color insert.)** (a) Wavelength and bandwidth of the selective reflection peak. (b) The reflection spectra. (c) The polarizing optical microscope textures under different amplitudes of the electric field. (Xiang, J. et al.: *Adv. Mater. 3014.* 2015. Copyright Wiley-VCH Verlag GmbH & Co. KGaA. Reproduced with permission.)

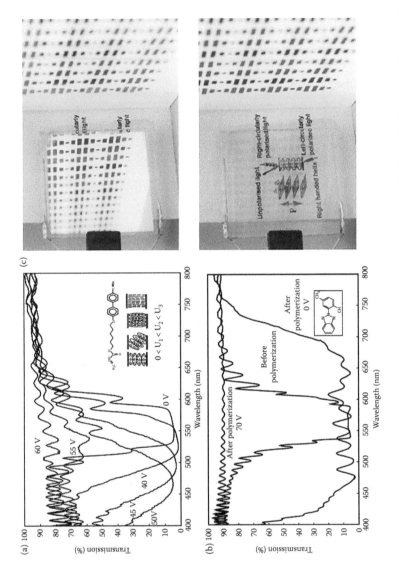

FIGURE 1.9 (a) The transmission spectra as a function of applied voltage of polymerization of a cholesteric system at zero voltage and the polymerized system at 70 V. (c) Photographs of an object reflected by a broadband reflector with the voltage off (top) and on (bottom). (Reprinted by permission from Macmillan Publishers Ltd. *Nature*, Hikmet, R. A. M.; Kemperman, H., 392, 476, copyright 1998.)

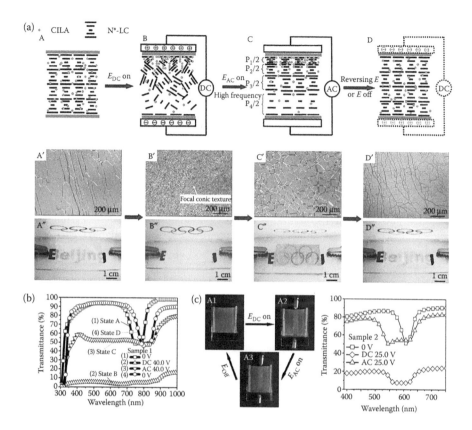

FIGURE 1.10 (a) Schematic representation of the molecular arrangements, microscopy images of the textures observed under polarized optical microscope and the corresponding digital photos of different electric field induced states. (b) The transmission spectra of the electrically switched CLC composite at different states. (c) Digital photos of the cells before and after patterns are addressed and the corresponding transmission spectra. (Hu, W. et al.: *Adv. Mater.* 468. 2010. Copyright Wiley-VCH Verlag GmbH & Co. KGaA. Reproduced with permission.)

reflecting visible light around it without the need of backlight and also promising to realize at low power consumption (Figure 1.10c).[58,60,65]

Thenceforth, there have been several great efforts focused on the electrically induced reflection bandwidth enhancement by deformation of polymer network in PSCLCs. For example, Bunning et al.[58] report on a PSCLC system prepared with negative dielectric anisotropy LC hosts, which exhibit large magnitude, reversible and repeatable symmetric broadening of the reflection notches under the control of electric field. As much as sevenfold increase in the reflection bandwidth can be achieved at 4 V/μm DC electric field enabling new applications such as dynamic high-efficiency filters, wave plates, and polarizers. White et al.[60] report a color-tunable mirrors based on electrically tunable bandwidth broadening of PSCLCs (Figure 1.11). The mechanism is the ion-facilitated electromechanical distortion of the polymer network (act as a dominant role of "structural chirality") resulting in a variation in pitch, hence, a broadening of the reflection bandwidth. With the DC voltage increasing, the selective reflection can be dynamically switched from selectively reflective (colored) into a broadband reflection (mirror), which is potentially useful in application areas such as displays, smart windows, and optical systems. Schenning et al.[65] develop an electrically tunable IR reflector with PSCLCs containing a negative dielectric, anisotropic LC, and a long flexible ethylene glycol twin cross-linker. Upon application of a small DC electric field, the reflection bandwidth can be broadened to unprecedented 1100 nm in the IR region, which has great potentials as smart windows of buildings and automobiles for energy saving.

CLC materials can also be precisely controllable by light through simple addition of light-driven molecular switches or motors such as azobenzenes, dithienylcyclopentenes, and spirooxazine derivatives,[66–68] which undergo light-driven pitch modulation and/or helix inversion upon light irradiation. Light-controllable CLCs exploit the benefits of light stimuli such as the ease of spatial, temporal, and remote control of the irradiation under ambient conditions.

Li et al.[69] designed and synthesized novel enantiomeric light-driven dithienylethene chiral switches with axial chirality, which transform from a colorless ring-open form to a colored ring-closed form (Figure 1.12a). As shown in Figure 1.12b, the two dithienylethene switches not only possess very high HTPs at their initial states but also exhibit a remarkable increase in HTP from the open to the closed form upon

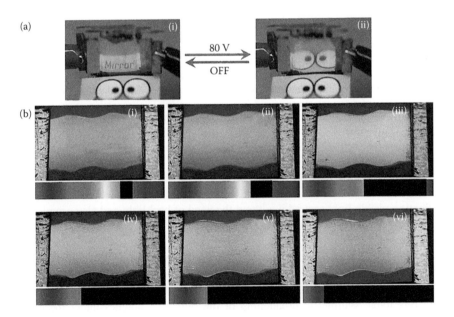

FIGURE 1.11 (See color insert.) (a) Photographs of the transmission and reflection transforming mode of the PSCLC at (i) 0 V and (ii) 80 V DC field. (b) The reflection-tunability of the PSCLC is visualized at (i) 0 V, (ii) 15 V, (iii) 30 V, (iv) 60 V, (v) 90 V, and (vi) 110 V. (Reprinted with permission from Lee, K. M. et al. *ACS Photonics* 1, 1033. Copyright 2014. American Chemical Society.)

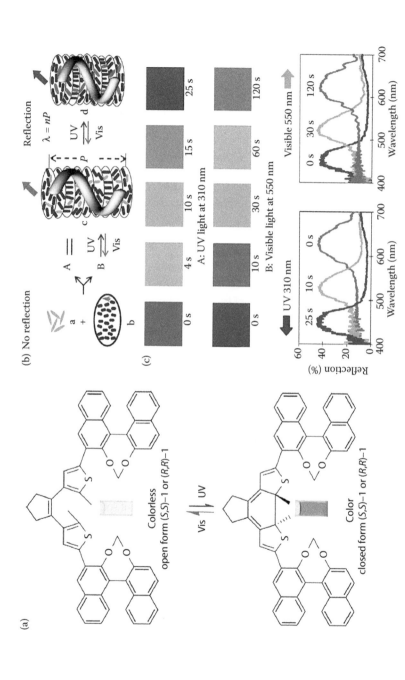

FIGURE 1.12 (a) Chemical structures and photoisomerization of dithienylcyclopentene chiral switches. (b) A schematic mechanism of the reflection wavelength of light-driven chiral molecular switch. (c) Reflective polarized microscope images and reflective spectra upon UV- and visible-light irradiation. (Reprinted with permission from Li, Y.; Urbas, A.; Li, Q. *J. Am. Chem. Soc.* 134, 9573. Copyright 2012 American Chemical Society.)

UV irradiation. Furthermore, the reverse isomerization process can be achieved simply by visible-light irradiation. Upon UV irradiation, the increase in HTP enables a hypsochromic shift of the refection bandgap from 630 nm (initial), 530 nm (10 s) to 440 nm (25 s), whereas visible-light irradiation results in a bathochromic shift of the refection bandgap with the corresponding refection images and the reflective spectra shown in Figure 1.12c. The numerous virtues of the light-controllable dithienylethene chiral switches such as high HTPs, reversible light tuning, superthermal stability make them promising candidates for practical technical applications, such as tunable color filters,[70–72] tunable mirrorless lasers,[73–75] and optically addressed displays[76–79] that require no driving electronics and can be flexible.

Yang et al.[80] proposed a PSCLC system composed of CLC, photopolymerizable monomer, and chiral azobenzene for broadband reflection. The chiral azobenzene binaphthyl molecular switch doped in CLCs undergoes reversible *trans–cis* isomerization upon light irradiation. The *trans*-isomer is linear and thermodynamically stable, whereas the *cis*-isomer is bent and thermodynamically metastable. The difference in molecular geometry of the *trans* and *cis* forms results in changes in the molecular conformation and HTP.[81] This mechanism of the formation of the broad reflection band uses the properties of the change in HTP of chiral azobenzene and its ability of UV absorption to induce the diffusion of reactive mesogens (Figure 1.13a). As shown in Figure 1.13b, upon increasing the irradiation time of 365 nm UV light, HTP of the doped chiral azobenzene decreases which mainly leads to a gradient distribution of pitch (Figure 1.13c) and enables a broadband reflection covering the wavelength from 1000 to 2400 nm (Figure 1.13d). Due to the anchoring effect of polymer network after polymerization, the PSCLC is hardly tuned by visible light (Figure 1.13d). This work suggests a new direction for designing broadband reflection and it is promising for the fabrication of broadband devices.

1.4 SURPASSING THE REFLECTANCE LIMITATION

Due to the polarization-selectivity rule of CLCs, the reflectance only reach 50% at best for unpolarized incident light. For improving the efficiency of reflectance, many researchers have drawn their inspiration from nature. For example, some members of the order *Coleoptera* exhibit colorful cuticle reflection by inherently fantastic structures,[82,83] among which golden beetle *Chrysina resplendens* can reflect both circular polarizations. Previous literature[84] represented the three-layer system of cuticle, in which a unidirectional layer was sandwiched between two LH helices (Figure 1.14a). The unidirectional layer plays an important role as half-wave plate for optical phase retardation. After passing through the cuticle, left-circularly polarized (LCP) light is directly reflected in the first layer and right-circularly polarized (RCP) light participate in a series of processes, as follow: (i) transmitted in the first layer, (ii) converted to LCP light through the second layer, (iii) reflected by the third layer, (iv) converted to RCP light through the second layer (after reflection from third layer), and (v) transmitted in the first layer. And hence, both left- and right-circular polarizations are totally reflected by this special sandwich structure. Several reports demonstrated man-made multilayer LC system succeeding in replicating the cuticle of *C. resplendens* for total reflectance (Figure 1.14b and c).[85–89] More classically, hyper-reflection

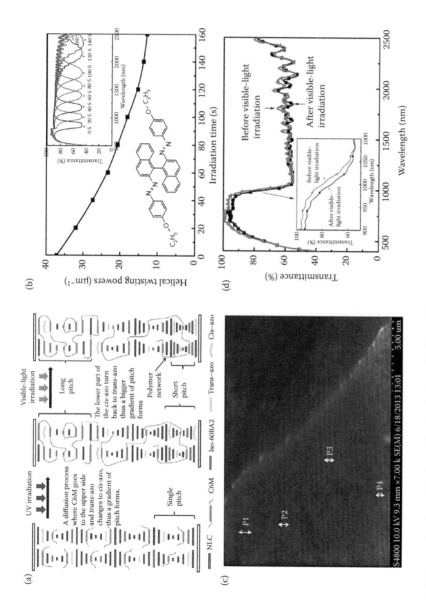

FIGURE 1.13 (a) The mechanism of the broadband reflection. (b) The HTPs change as the irradiation time increases, the insert is the change of transmittance spectra as the irradiation time increases and the structure of the doped chiral azobenzene 2C. *(Continued)*

FIGURE 1.13 (Continued)   (c) The SEM picture of pitch distribution. (d) The transmittance spectra of the PSCLC with a broadband reflection from 1000 to 2400 nm before and after visible-light irradiation. (Chen, X. et al. *Chem. Com.* 2014, 50, 691. Reproduced by permission of The Royal Society of Chemistry.)

can be realized by stacking CLC layers with similar pitch but opposite helicity sense for the specific functions.[85,90–93] The advantage of multilayer solution is the layer-by-layer replicability. However, reducing the interlayer defects, accompanying with the optical losses, and enhancing the efficiency of implementation are challenges for future investigation.[94]

Mitov et al.[95–98] aimed to fabricate a single-layer PSCLC film with both handed helices coexisting in the IR spectrum. As is shown in Figure 1.15a, the handedness inversion of CLC is induced by a thermal-responsive chiral dopant, the mesogenic ester (S)-1,2-propanediol, at the critical temperature T_c. Thus, the PSCLC is prepared when the helical sense is RH with pitch p_0 at temperature T_2 and then a portion of RH CLCs can be stabilized around the polymer network, which also acts as an

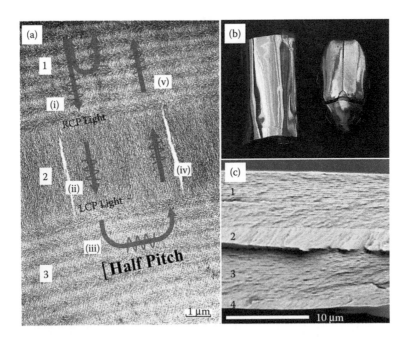

FIGURE 1.14   (a) TEM photograph of the fractured surface of cuticle of *Chrysina resplendens*, including three layers: (1) left-handed helix, (2) unidirectional layer, and (3) left-handed helix. (Caveney, S. *Proc. Roy. Soc. Lond. B* 1971, 178, 205. Reproduced by permission of The Royal Society of Chemistry.) Schematic mechanism of light path is shown in the insert (i–v). (b) Photograph of an LC composite polymer (left) and a *Chrysina resplendens* beetle (right). (c) SEM photograph of the LC composite film, including four layers: (1) cholesteric, (2) untwisted nematic retarder layer, (3) cholesteric, and (4) untwisted alignment layer. (Matranga, A. et al. *Adv. Mater.* 520. 2013. Copyright Wiley-VCH Verlag GmbH & Co. KGaA. Reproduced with permission.)

internal memory effect of the RH helix characteristics by itself (Figure 1.15d and e). While cooling to temperature T_1, LH helix with the same pitch p_0 reoccurs after the handedness inversion (Figure 1.15b).[99] The unpolarized, RCP and LCP transmittance spectra of the PSCLC demonstrate that the reflectance go beyond the typical limitation of 50% in all polarization states (Figure 1.15c).[95] This technique firstly offers a novel self-organized composite system in a single layer with a double helical handedness. However, due to the weak HTP of the dopant, the reflective wavelength is only achieved in IR region, which restricts to the applied areas of devices.[100]

Yang et al.[101–105] adopted "washout/refill" techniques to fabricate an original structure combining with the memory effect of polymer network and the postincoming LCs for different requirements. As the prepared procedures in Figure 1.16a, a helical

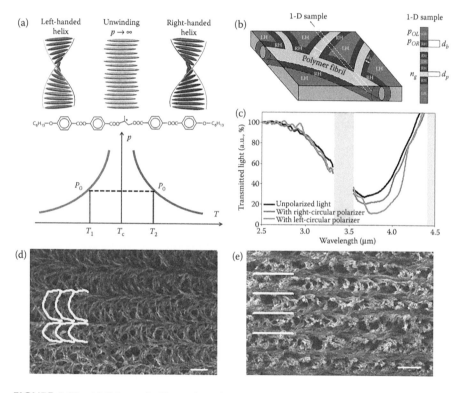

FIGURE 1.15 (a) Schematic illustration of the thermally induced handedness inversion of CLC, chemical structure of the chiral ester (S)-1,2-propanediol and the change plot of the pitch length of CLC (p) as a function of temperature (T). (b) Schematic model adopted for the PSCLC, where n_g is the polymer's refractive index, d_p is the thickness of the polymer fibrils, d_b is the thickness of RH CLC bound to the polymer network, and P_{OL}, P_{OR} are the pitch lengths for the LH and RH regions, respectively. (Reprinted from *Opt. Commun.*, 282, Tasolamprou, A. C. et al., 903, Copyright 2009, with permission from Elsevier.) (c) The unpolarized, RCP and LCP transmittance spectra of the PSCLC. The shadowed areas mark the chemical absorption bands. (d, e) SEM images of the polymer network in the view of surface and cross section, respectively. (Reprinted by permission from Macmillan Publishers Ltd. *Nat. Mater.*, Mitov, M.; Dessaud, N., 5, 361, copyright 2006.)

polymer network was formed in an LH CLC mixture at the beginning (Figure 1.16b, curve 2, and Figure 1.16c). Then the low-molar-mass nonreactive CLCs were removed from the mesh of polymer network by immersing the sample in cyclohexane and tetrahydrofuran (Figure 1.16b, curve 3). Thereafter, RH CLC with the same pitch as polymer network (Figure 1.16b, curve 1) was refilled into the free fraction of polymer network. Using this mechanism, the transmittance of unpolarized light reduces from 50% to nearly 0%, or in other words, reflection increases close to 100% (Figure 1.16b, curve 4). By this strategy in further investigation, the postincoming CLCs are relatively easier to be functional to external stimuli in a suitable network. Upon treatment with electric field,[105–107] temperature control[104,108–112] or light,[113,114] the broadwidth-, multiple-, and hyper-reflection bandgap can be accomplished by dynamic modulation. Such a method has a great potential of various applications, for example, lasing emission,[115] smart windows,[107] tunable reflective display,[106] etc. Even so, the memory effect of polymer template needs to be employed on a settled

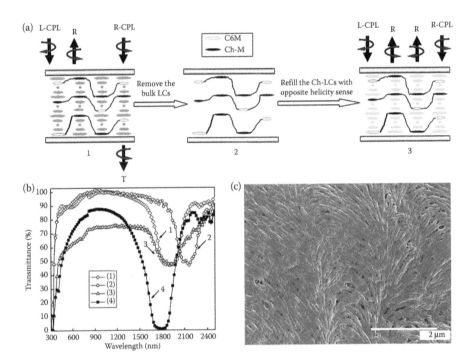

FIGURE 1.16 (a) Schematic illustration of "washout/refill" strategy for composite CLC films with both RCP and LCP reflection bands. Step 1: The LH PSCLC is prepared in a given pitch by UV-light polymerization. Step 2: The nonreactive LCs are removed by dissolving into the proper solvent. Step 3: Another low-molar-mass CLC with the same given pitch in RH helix is introduced into the porous LH polymer network. (b) The transmittance spectra of (1) the prefilled RH CLC, (2) the photopolymerizable CLC before UV exposure and (3) after UV exposure, and (4) the cell with the polymer network after refilling the RHCLC. (c) SEM image of the helical polymer network. (Reprinted with permission from Guo, J. et al. *Appl. Phys. Lett.* 2008, 93, 201901. Copyright 2008, American Institute of Physics.)

surface with at least three steps, which is the key problem for prompting the devices in prospective field.

Interestingly, in order to explore a novel design for the hyper-reflection of CLC in a convenient and controllable approach, Yang et al.[116] have designed an optically tunable self-organized helical superstructures in a single-layer system by doping photo-responsive molecular motor. Due to the handedness inversion of overcrowded alkene M5 upon UV-light irradiation (Figure 1.17b insert), accompanying by the photoisomerization from stable form P to unstable form M,[117,118] it was found that the UV dependence of the photo-equilibrium HTP value (β_{PSS}), corresponding to the photo-changed molar ratio of unstable form M at the PSS (χ_{PSS}), with a rapid respond time in Figure 1.17a and b. On the basis of the UV absorption ability of M5 in accordance with the Beer–Lambert law,[119] the gradient distribution of UV intensity occurs across a thickness of the CLC film. By dominating the different distribution of χ_{PSS} with the UV-intensity damping in different levels along the helical axis, the nonuniform arrangements of helical pitch with the same or opposite handedness is shown in Figure 1.17c. While 35-μm CLC exposure to 74.0 mW cm^{-2} UV light, the unpolarized, RCP and LCP transmittance spectra indicate all the polarized reflectance break through the limitation of 50% at that moment, in which the helical pitches of the LH and RH CLC are approximated (Figure 1.17d). To investigate with SEM, the self-organized distribution of helical pitch was verified from top to bottom in the PSCLC, $p_1 < p_2 < p_3 > p_4 > p_5$, distinguishing from the coexisting CLCs with the opposite handedness in the abovementioned ways (Figure 1.17e). However, this photo-tuning composite system displays a reversibly addressable optical performances switching in not only hyper-reflection even dynamic single-bandgap shifting and broad-bandgap reflection in a single-layer film. As a result, this technique can plainly overcome the selective reflection rule of CLCs (bandwidth or reflectance) by light stimulus, instead of layer-by-layer stacking, thermo/photoinduced diffusional solid elastomer, generating memory effects or inorganic/organic template.[37]

1.5 CONCLUSION AND OUTLOOK

CLC materials are witnessing a significant surge in interest because of their unique property of selective reflection of light in the desired wavelength range and high controllable performance to the external stimulus. This chapter has demonstrated the recent progress of novel CLC materials and innovative cholesteric-liquid-crystalline architectures with broad reflection bandwidth and high reflectance beyond 50%, and several effective methods and processing technique have been developed in the past decades. In summary, both reflection limitations of CLCs have been exceeded by the abovementioned experimental strategies through physical, chemical methodology, or combination of both. To the subject of CLCs with a broad reflection band, a variation of pitch is integral, whether a pitch gradient or a random distribution in the volume of materials. And to obtain a hyper-reflection band with both circularly polarized states, simultaneous LH and RH helical structures consist in the hybrid system, which is layer-by-layer arrangement or interpenetrating materials. In addition, the choice of a specific procedure, or combination of procedures, strongly depends on the fundamental question to be addressed or depending on the desired application.

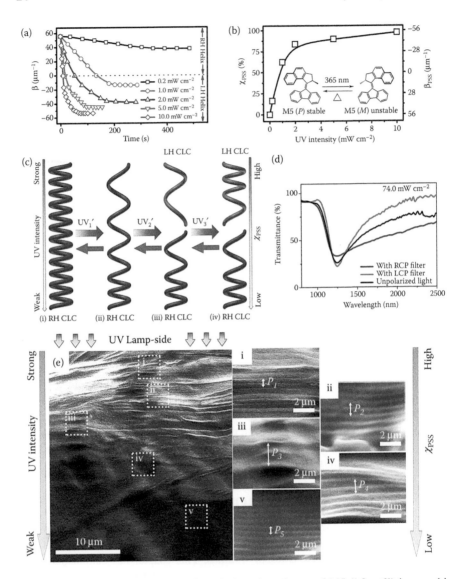

FIGURE 1.17 (a) The HTP value, β, variation of a mixture of M5 (1.0 wt%) in an achiral nematic LC host at different UV intensity by Grandjean-Cano wedge method. (b) The change plot of χ_{PSS} and β_{PSS} as a function of UV intensity. The molecular isomerization of overcrowded alkene M5 under UV exposure is shown in the insert. (c) Schematic illustrations of the photodynamic CLC molecular arrangements in a thicker cell with different intensity $UV_1' < UV_2' < UV_3'$: (i) The homogeneous helical pitch of CLC at the initial state; (ii–iv) The three different possible distribution of helical pitch of CLC with the same or opposite handedness, corresponding to the gradient-distribution χ_{PSS} in different degrees. (d) The unpolarized, RCP and LCP transmittance spectra of CLC mixture driven with 74.0 mW cm⁻² UV irradiation in 35 μm cell. (e) SEM image of the fractured surface of the PSCLC. (i–v) Part regions under a higher resolution. The organized distribution of pitch with an increased degrees at the upper side and a gradual decreasing at the lower side, $p_1 < p_2 < p_3 > p_4 > p_5$.

Wide-band CLC materials are expected to have great potential applications in fields such as reflective colored display, brightness-enhancement films, or smart switchable windows for energy efficiency.[120] For example, CLC polarizers with broadband reflection over the entire visible spectrum can greatly improve the light yield and energy efficiency of liquid-crystal devices by reflecting and recycling the opposite polarized light in the backlight system.[22] Moreover, a CLC reflector that is transparent to visible light and highly reflective for IR light is expected to be applied in smart switchable reflective window to control solar light and heat. Beyond the low-energy-consumption process, IR CLC reflectors block a considerable amount of incident, unwanted heat and therefore a significant amount of energy could be saved on cooling especially in summer.[94,121] In addition, CLCs represent fascinating prospect in laser protection owing to their inherent self-organized periodic helical superstructures within a "green" approach. The wavelength of CLC reflection can be accurately modulated with the doped dye absorption[122] to obtain the maximum synergistic effect for optimal optical density (OD), providing a high-performance protection with tunable wavelength, feasible fabrication, high OD, and good visibility.[123] Moreover, employing CLCs polymer films as transparent and flexible back-reflectors for dye-sensitized solar cells (DSCs) has been proved to be a facile strategy to enhance the optical efficiency of the devices. The selective light reflection of these CLC films gives rise to the possibility for increasing the optical path length of the light in particular wavelength region while retaining the cell transparency. The enhancement of photocurrent and power conversion efficiency (PCE) reveals strong wavelength dependence owing to the selective reflection of these CLC polymer films. The DSCs with proper combination of CLC back-reflectors yield the maximum enhancement over 21% in photocurrent and 17% in PCE.[124]

These CLC materials have enormous perspectives not only in everyday life but also in advanced or challenging applications including artificial photonic circuits or integrated optics and fiber-optic communications. Actually there is still much work to be done in making innovations in CLC materials with excellent optical performance, highly sensitivity, and reversibility to external stimuli. Meanwhile, developing preparation technologies and procedures for industrial production are crucial for the practical application.

REFERENCES

1. Reinitzer, F. Beiträge zur kenntniss des cholesterins. *Monatsh. Chem.* 1888, 9, 421.
2. Small, D. M. The physical state of lipids of biological importance: Cholesteryl esters, cholesterol, triglyceride. In *Surface Chemistry of Biological Systems* (Ed. Blank, M.); Springer: New York, 1970, p. 55.
3. Sharma, V., Crne, M., Park, J. O., Srinivasarao, M. Structural origin of circularly polarized iridescence in jeweled beetles. *Science* 2009, 325, 449.
4. Hamley, I. W. Liquid crystal phase formation by biopolymers. *Soft Matter* 2010, 6, 1863.
5. Rey, A. D. Liquid crystal models of biological materials and processes. *Soft Matter* 2010, 6, 3402.
6. Pieraccini, S., Masiero, S., Ferrarini, A., Spada, G. P. Chirality transfer across length-scales in nematic liquid crystals: Fundamentals and applications. *Chem. Soc. Rev.* 2011, 40, 258.

7. Li, Y., Li, Q. Photoresponsive cholesteric liquid crystals. In *Intelligent Stimuli-Responsive Materials: From Well-Defined Nanostructures to Applications* (Ed. Li, Q.); John Wiley & Sons: Hoboken, 2013, p. 141.

8. Li, Y., Li, Q. Photoresponsive chiral liquid crystal materials: From 1D helical super-structures to 3D periodic cubic lattices and beyond. In *Nanoscience with Liquid Crystals: From Self-Organized Nanostructures to Applications* (Ed. Li, Q.); Springer: Heidelberg, 2014, p. 135.

9. Wang, L. Self-activating liquid crystal devices for smart laser protection. *Liq. Cryst.* 2016, 43, 2062.

10. Broer, D. J. Polymerization mechanisms. In *Radiation Curing in Polymer Science and Technology: Practical Aspects and Applications* (Eds. Fouassier, J. P., Rabek, J. F.); Elsevier: Amsterdam, 1993; p. 383.

11. Broer, D. J., Heynderickx, I. Three-dimensionally ordered polymer networks with a helicoidal structure. *Macromolecules* 1990, 23, 2474.

12. Broer, D. J., Lub, J., Mol, G. N. Photo-controlled diffusion in reacting liquid crystals: A new tool for the creation of complex molecular architectures. *Macromol. Symp.* 1997, 117, 33.

13. Broer, D. J. Deformed chiral-nematic networks obtained by polarized excitation of a dichroic photoinitiator. *Curr. Opin. Solid St. M.* 2002, 6, 553.

14. Lub, J., Broer, D. J., van de Witte, P. Colourful photo-curable coatings for application in the electro-optical industry. *Prog. Org. Coat.* 2002, 45, 211.

15. Leewis, C. M., de Jong, A. M., van Ijzendoorn, L. J., Broer, D. J. Reaction-diffusion model for the preparation of polymer gratings by patterned ultraviolet illumination. *J. Appl. Phys.* 2004, 95, 4125.

16. Broer, D. J., Finkelmann, H., Kondo, K. In-situ photopolymerization of oriented liquid-crystalline acrylate, 1. Preservation of molecular order during photopolymerization. *Makromol. Chem.* 1988, 189, 185.

17. Broer, D. J., Mol, G. N., Challa, G. In-situ photopolymerization of oriented liquid-crystalline acrylate, 2. Kinetic aspects of photopolymerization in the mesophase. *Makromol. Chem.* 1989, 190, 19.

18. Broer, D. J., Boven, J., Mol, G. N., Challa, G. In-situ photopolymerization of oriented liquid-crystalline acrylates, 3. Oriented polymer networks from a mesogenic diacrylate. *Makromol. Chem.* 1989, 190, 2255.

19. Broer, D. J., Hikmet, R. A. M., Challa, G. In-situ photopolymerization of oriented liquid-crystalline acrylates, 4. Influence of a lateral methyl substituent on monomer and oriented polymer network properties of a mesogenic diacrylate. *Makromol. Chem.* 1989, 190, 3201.

20. Broer, D. J., Mol, G. N., Challa, G. In-situ photopolymerization of oriented liquid-crystalline acrylates, 5. Influence of the alkylene spacer on the properties of the meso-genic monomers and the formation and properties of oriented polymer networks. *Makromol. Chem.* 1991, 192, 59.

21. Lub, J., Broer, D. J., Hikmet, R. A. M., Nierop, K. G. J. Synthesis and photopolymeriza-tion of cholesteric liquid-crystalline diacrylates. *Liq. Cryst.* 1995, 18, 319.

22. Broer, D. J., Lub, J., Mol, G. N. Wide-band reflective polarizers from cholesteric poly-mer networks with a pitch gradient. *Nature* 1995, 378, 467.

23. Broer, D. J., Mol, G. N., van Haaren, J., Lub, J. Photo-induced diffusion in polymer-izing chiral-nematic media. *Adv. Mater.* 1999, 11, 573.

24. Leewis, C. M., Simons, D. P. L., de Jong, A. M., Broer, D. J., de Voigt, M. J. A. PIXE monitoring of diffusion during photo-polymerization. *Nucl. Instrum. Meth. B* 2000, 161, 651.

25. Lub, J., Broer, D. J., Wegh, R. T., Peeters, E., van der Zande, B. M. I. Formation of optical films by photo-polymerisation of liquid crystalline acrylates and

application of these films in liquid crystal display technology. *Mol. Cryst. Liq. Cryst.* 2005, 429, 77.

26. Relaix, S., Bourgerette, C., Mitov, M. Broadband reflective liquid crystalline gels due to the ultraviolet light screening made by the liquid crystal. *Appl. Phys. Lett.* 2006, 89, 291507.

27. Relaix, S., Bourgerette, C., Mitov, M. Broadband reflective cholesteric liquid crystal-line gels: Volume distribution of reflection properties and polymer network in relation with the geometry of the cell photopolymerization. *Liq. Cryst.* 2007, 34, 1009.

28. Fan, B., Vartak, S., Eakin, J. N., Faris, S. M. Broadband polarizing films by photopolymerization-induced phase separation and in situ swelling. *Appl. Phys. Lett.* 2008, 92, 061101.

29. Fan, B., Vartak, S., Eakin, J. N., Faris, S. M. Surface anchoring effects on spectral broadening of cholesteric liquid crystal films. *J. Appl. Phys.* 2008, 104, 023108.

30. Zhang, L. P., Li, K. X., Hu, W., Cao, H., Cheng, Z. H., He, W. L., Xiao, J. M., Yang, H. Broadband reflection mechanism of polymer stabilised cholesteric liquid crystal (PSChLC) with pitch gradient. *Liq. Cryst.* 2011, 38, 673.

31. Mitov, M., Boudet, A., Sopena, P. From selective to wide-band light reflection: A sim-ple thermal diffusion in a glassy cholesteric liquid crystal. *Eur. Phys. J. B* 1999, 8, 327.

32. Binet, C., Mitov, M., Boudet, A. Bragg reflections in cholesteric liquid crystals: From selectivity to broadening and reciprocally. *Mol. Cryst. Liq. Cryst.* 2000, 339, 111.

33. Boudet, A., Binet, C., Mitov, M., Bourgerette, C., Boucher, E. Microstructure of vari-able pitch cholesteric films and its relationship with the optical properties. *Eur. Phys. J. E* 2000, 2, 247.

34. Mitov, M., Binet, C., Boudet, A., Bourgerette, C. Glassy cholesteric broadband reflec-tors with a pitch gradient: Material design, optical properties and microstructure. *Mol. Cryst. Liq. Cryst.* 2001, 358, 209.

35. Lavernhe, A., Mitov, M., Binet, C., Bourgerette, C. How to broaden the light reflec-tion band in cholesteric liquid crystals? A new approach based on polymorphism. *Liq. Cryst.* 2001, 28, 803.

36. Zografopoulos, D. C., Kriezis, E. E., Mitov, M., Binet, C. Theoretical and experimental optical studies of cholesteric liquid crystal films with thermally induced pitch gradi-ents. *Phys. Rev. E* 2006, 73, 061701.

37. Mitov, M. Cholesteric liquid crystals with a broad light reflection band. *Adv. Mater.* 2012, 24, 6260.

38. Xiao, J., Cao, H., He, W., Ma, Z., Geng, J., Wang, L., Wang, G., Yang, H. Wide-band reflective polarizers from cholesteric liquid crystals with stable optical properties. *J. Appl. Polym. Sci.* 2007, 105, 2973.

39. Guo, J. B., Sun, J., Zhang, L. P., Li, K. X., Cao, H., Yang, H., Zhu, S. Q. Broadband reflection in polymer stabilized cholesteric liquid crystal cells with chiral monomers derived from cholesterol. *Polym. Advan. Technol.* 2008, 19, 1504.

40. Guo, J. B., Liu, F., Zhang, L. P., Cao, H., Yang, H. Preparation and reflectance prop-erties of new cholesteric liquid crystalline copolymers containing cholesteryl group. *Polym. Eng. Sci.* 2009, 49, 937.

41. Sixou, P., Gautier, C., Guillard, H. Passive broadband reflector: Elaboration and spec-tral properties. *Mol. Cryst. Liq. Cryst.* 2001, 364, 665.

42. Sixou, P., Gautier, C. Passive broadband reflector using photocrosslinkable liquid crys-tal molecules. *Polym. Advan. Technol.* 2002, 13, 329.

43. Bian, Z., Li, K., Huang, W., Cao, H., Yang, H., Zhang, H. Characteristics of selective reflection of chiral nematic liquid crystalline gels with a nonuniform pitch distribution. *Appl. Phys. Lett.* 2007, 91, 201908.

44. Huang, W., Bian, Z. Y., Li, K. X., Xiao, J. M., Cao, H., Yang, H. Study on selective reflection properties of chiral nematic liquid crystalline composites with a non-uniform pitch distribution. *Liq. Cryst.* 2008, 35, 1313.

45. Xiao, J., Zhao, D., Cao, H., Yang, H. New micro-structure designs of a wide band reflective polarizer with a pitch gradient. *Liq. Cryst.* 2007, 34, 473.
46. Gao, Y. Z., Yao, W. H., Sun, J., Zhang, H. M., Wang, Z. D., Wang, L., Yang, D. K., Zhang, L. Y., Yang, H. A novel soft matter composite material for energy-saving smart windows: From preparation to device application. *J. Mater. Chem. A* 2015, 3, 10738.
47. Zhang, L., Wang, M., Wang, L., Yang, D.-k., Yu, H., Yang, H. Polymeric infrared reflective thin films with ultra-broad bandwidth. *Liq. Cryst.* 2016, 43, 750.
48. Collings, P. J. *Liquid Crystals: Nature's Delicate Phase of Matter*; Princeton University Press: New Jersey, 2002.
49. Kikuchi, H., Kibe, S., Kajiyama, T. Sharp steepness of molecular reorientation for nematics containing liquid crystalline polymer. *Proc. SPIE* 1995, 2408, 141.
50. Yang, H., Yamane, H., Kikuchi, H., Yamane, H., Zhang, G., Chen, X. F., Tisato, K. Investigation of the electrothermo-optical effect of a smectic LCP-nematic LC-chiral dopant ternary composite system based on $S_A \leftrightarrow N^*$ phase transition. *J. Appl. Polym. Sci.* 1999, 73, 623.
51. Wang, F. F., Cao, H., Li, K. X., Song, P., Wu, X. J., Yang, H. Control homogeneous alignment of chiral nematic liquid crystal with smectic-like short-range order by thermal treatment. *Colloid Surf. A-Physicochem. Eng. Asp.* 2012, 410, 31.
52. Lin, T. H., Jau, H. C., Chen, C. H., Chen, Y. J., Wei, T. H., Chen, C. W., Fuh, A. Y. G. Electrically controllable laser based on cholesteric liquid crystal with negative dielectric anisotropy. *Appl. Phys. Lett.* 2006, 88, 061122.
53. Yang, H., Mishima, K., Matsuyama, K., Hayashi, K.-I., Kikuchi, H., Kajiyama, T. Thermally bandwidth-controllable reflective polarizers from (polymer network/liquid crystal/chiral dopant) composites. *Appl. Phys. Lett.* 2003, 82, 2407.
54. Hrozhyk, U. A., Serak, S. V., Tabiryan, N. V., Bunning, T. J. Optical tuning of the reflection of cholesterics doped with azobenzene liquid crystals. *Adv. Funct. Mater.* 2007, 17, 1735.
55. Choi, S. S., Morris, S. M., Huck, W. T. S., Coles, H. J. The switching properties of chiral nematic liquid crystals using electrically commanded surfaces. *Soft Matter* 2009, 5, 354.
56. Li, C. C., Tseng, H. Y., Pai, T. W., Wu, Y. C., Hsu, W. H., Jau, H. C., Chen, C. W., Lin, T. H. Bistable cholesteric liquid crystal light shutter with multielectrode driving. *Appl. Opt.* 2014, 53, E33.
57. McConney, M. E., Tondiglia, V. P., Natarajan, L. V., Lee, K. M., White, T. J., Bunning, T. J. Electrically induced color changes in polymer-stabilized cholesteric liquid crystals. *Adv. Opt. Mater.* 2013, 1, 417.
58. Tondiglia, V. T., Natarajan, L. V., Bailey, C. A., Duning, M. M., Sutherland, R. L., Ke-Yang, D., Voevodin, A., White, T. J., Bunning, T. J. Electrically induced bandwidth broadening in polymer stabilized cholesteric liquid crystals. *J. Appl. Phys.* 2011, 110, 053109.
59. Tondiglia, V. P., Natarajan, L. V., Bailey, C. A., McConney, M. E., Lee, K. M., Bunning, T. J., Zola, R., Nemati, H., Yang, D.-K., White, T. J. Bandwidth broadening induced by ionic interactions in polymer stabilized cholesteric liquid crystals. *Opt. Mater. Express* 2014, 4, 1465.
60. Lee, K. M., Tondiglia, V. P., McConney, M. E., Natarajan, L. V., Bunning, T. J., White, T. J. Color-tunable mirrors based on electrically regulated bandwidth broadening in polymer-stabilized cholesteric liquid crystals. *ACS Photonics* 2014, 1, 1033.
61. Nemati, H., Liu, S., Zola, R. S., Tondiglia, V. P., Lee, K. M., White, T., Bunning, T., Yang, D. K. Mechanism of electrically induced photonic band gap broadening in polymer stabilized cholesteric liquid crystals with negative dielectric anisotropies. *Soft Matter* 2015, 11, 1208.

62. Xiang, J., Li, Y., Li, Q., Paterson, D. A., Storey, J. M. D., Imrie, C. T., Lavrentovich, O. D. Electrically tunable selective reflection of light from ultraviolet to visible and infrared by heliconical cholesterics. *Adv. Mater.* 2015, 27, 3014.

63. Hikmet, R. A. M., Kemperman, H. Electrically switchable mirrors and optical components made from liquid-crystal gels. *Nature* 1998, 392, 476.

64. Hu, W., Zhao, H., Song, L., Yang, Z., Cao, H., Cheng, Z., Liu, Q., Yang, H. Electrically controllable selective reflection of chiral nematic liquid crystal/chiral ionic liquid composites. *Adv. Mater.* 2010, 22, 468.

65. Khandelwal, H., Debije, M. G., White, T. J., Schenning, A. Electrically tunable infrared reflector with adjustable bandwidth broadening up to 1100 nm. *J. Mater. Chem. A* 2016, 4, 6064.

66. Li, Q., Li, L. F., Kim, J., Park, H. S., Williams, J. Reversible photoresponsive chiral liquid crystals containing a cholesteryl moiety and azobenzene linker. *Chem. Mater.* 2005, 17, 6018.

67. Li, Y. N., Urbas, A., Li, Q. Synthesis and characterization of light-driven dithienylcyclopentene switches with axial chirality. *J. Org. Chem.* 2011, 76, 7148.

68. Jin, L. M., Li, Y. N., Ma, J., Li, Q. A. Synthesis of novel thermally reversible photochromic axially chiral spirooxazines. *Org. Lett.* 2010, 12, 3552.

69. Li, Y., Urbas, A., Li, Q. Reversible light-directed red, green, and blue reflection with thermal stability enabled by a self-organized helical superstructure. *J. Am. Chem. Soc.* 2012, 134, 9573.

70. Hochbaum, A., Jiang, Y., Li, L., Vartak, S., Faris, S. Cholesteric color filters: Optical characteristics, light recycling, and brightness enhancement. *SID Digest* 1999, 30, 1063.

71. Lub, J., van de Witte, P., Doornkamp, C., Vogels, J. P. A., Wegh, R. T. Stable photopatterned cholesteric layers made by photoisomerization and subsequent photopolymerization for use as color filters in liquid-crystal displays. *Adv. Mater.* 2003, 15, 1420.

72. Ha, N. Y., Ohtsuka, Y., Jeong, S. M., Nishimura, S., Suzaki, G., Takanishi, Y., Ishikaw, K., Takezoe, H. Fabrication of a simultaneous red-green-blue reflector using single-pitched cholesteric liquid crystal. *Nat. Mater.* 2008, 7, 43.

73. Kopp, V. I., Fan, B., Vithana, H. K. M., Genack, A. Z. Low-threshold lasing at the edge of a photonic stop band in cholesteric liquid crystals. *Opt. Lett.* 1998, 23, 1707.

74. Cao, W., Mu oz, A., Palffy-Muhoray, P., Taheri, B. Lasing in a three-dimensional photonic crystal of the liquid crystal blue phase II. *Nat. Mater.* 2002, 1, 111.

75. Furumi, S., Tamaoki, N. Glass-forming cholesteric liquid crystal oligomers for new tunable solid-state laser. *Adv. Mater.* 2010, 22, 886.

76. Murata, H., Mori, Y., Yamashita, S., Maenaka, A., Okada, S., Oyamada, K., Kishimoto, S. A real-time 2-D to 3-D image conversion technique using computed image depth. *SID Digest* 1998, 29, 919.

77. Tamaoki, N., Song, S., Moriyama, M., Matsuda, H. Rewritable full-color recording in a photon mode. *Adv. Mater.* 2000, 12, 94.

78. Venkataraman, N., Magyar, G., Lightfoot, M., Montbach, E., Khan, A., Schneider, T., Doane, J. W., Green, L., Li, Q. Thin flexible photosensitive cholesteric displays. *J. Soc. Inf. Display* 2009, 17, 869.

79. Li, Q., Li, Y., Ma, J., Yang, D.-K., White, T. J., Bunning, T. J. Directing dynamic control of red, green, and blue reflection enabled by a light-driven self-organized helical superstructure. *Adv. Mater.* 2011, 23, 5069.

80. Chen, X., Wang, L., Chen, Y., Li, C., Hou, G., Liu, X., Zhang, X., He, W., Yang, H. Broadband reflection of polymer-stabilized chiral nematic liquid crystals induced by a chiral azobenzene compound. *Chem. Commun.* 2014, 50, 691.

81. White, T. J., Bricker, R. L., Natarajan, L. V., Tabiryan, N. V., Green, L., Li, Q., Bunning, T. J. Phototunable azobenzene cholesteric liquid crystals with 2000 nm range. *Adv. Funct. Mater.* 2009, 19, 3484.

82. Michelson, A. A. LXI. On metallic colouring in birds and insects. *Philos. Mag. Ser. 6* 1911, 21, 554.

83. Seago, A. E., Brady, P., Vigneron, J.-P., Schultz, T. D. Gold bugs and beyond: A review of iridescence and structural colour mechanisms in beetles (Coleoptera). *J. R. Soc. Interface* 2009, 6, S165.

84. Caveney, S. Cuticle reflectivity and optical activity in scarab beetles: The role of uric acid. *Proc. Roy. Soc. Lond. B* 1971, 178, 205.

85. Makow, D. M. Peak reflectance and color gamut of superimposed left- and right-handed cholesteric liquid crystals. *Appl. Opt.* 1980, 19, 1274.

86. Song, M. H., Park, B. C., Shin, K. C., Ohta, T., Tsunoda, Y., Hoshi, H. Takanishi, Y. et al. Effect of phase retardation on defect-mode lasing in polymeric cholesteric liquid crystals. *Adv. Mater.* 2004, 16, 779.

87. Song, M. H., Shin, K. C., Park, B., Takanishi, Y., Ishikawa, K., Watanabe, J., Nishimura, S. et al. Polarization characteristics of phase retardation defect mode lasing in polymeric cholesteric liquid crystals. *Sci. Technol. Adv. Mat.* 2004, 5, 437.

88. Jisoo, H., Myoung hoon, S., Byoungchoo, P., Nishimura, S., Toyooka, T., Wu, J. W., Takanishi, Y., Ishikawa, K., Takezoe, H. Electro-tunable optical diode based on photonic bandgap liquid-crystal heterojunctions. *Nat. Mater.* 2005, 4, 383.

89. Matranga, A., Baig, S., Boland, J., Newton, C., Taphouse, T., Wells, G., Kitson, S. Biomimetic reflectors fabricated using self-organising, self-aligning liquid crystal polymers. *Adv. Mater.* 2013, 25, 520.

90. Gevorgyan, A. A., Papoyan, K. V., Pikichyan, O. V. Reflection and transmission of light by cholesteric liquid crystal-glass-cholesteric liquid crystal and cholesteric liquid crystal(1)-cholesteric crystal(2) systems. *Opt. Spektrosk.* 2000, 88, 586.

91. Song, M. H., Ha, N. Y., Amemiya, K., Park, B., Takanishi, Y., Ishikawa, K., Wu, J. W., Nishimura, S., Toyooka, T., Takezoe, H. Defect-mode lasing with lowered threshold in a three-layered hetero-cholesteric liquid-crystal structure. *Adv. Mater.* 2006, 18, 193.

92. Ha, N. Y., Jeong, S. M., Nishimura, S., Takezoe, H. Polarization-independent multiple selective reflections from bichiral liquid crystal films. *Appl. Phys. Lett.* 2010, 96, 153301.

93. Chen, G., Wang, L., Wang, Q., Sun, J., Song, P., Chen, X., Liu, X. et al. Photoinduced hyper-reflective laminated liquid crystal film with simultaneous multicolor reflection. *ACS Appl. Mater. Interfaces* 2014, 6, 1380.

94. Khandelwal, H., Loonen, R. C. G. M., Hensen, J. L. M., Schenning, A. P. H. J., Debije, M. G. Application of broadband infrared reflector based on cholesteric liquid crystal polymer bilayer film to windows and its impact on reducing the energy consumption in buildings. *J. Mater. Chem. A* 2014, 2, 14622.

95. Mitov, M., Dessaud, N. Going beyond the reflectance limit of cholesteric liquid crystals. *Nat. Mater.* 2006, 5, 361.

96. Mitov, M., Dessaud, N. Cholesteric liquid crystalline materials reflecting more than 50% of unpolarized incident light intensity. *Liq. Cryst.* 2007, 34, 183.

97. Relaix, S., Mitov, M. The effect of geometric and electric constraints on the performance of polymer-stabilized cholesteric liquid crystals with a double-handed circularly polarized light reflection band. *J. Appl. Phys.* 2008, 104.

98. Agez, G., Mitov, M. Cholesteric liquid crystalline materials with a dual circularly polarized light reflection band fixed at room temperature. *J. Phys. Chem. B* 2011, 115, 6421.

99. Tasolamprou, A. C., Mitov, M., Zografopoulos, D. C., Kriezis, E. E. Theoretical and experimental studies of hyperreflective polymer-network cholesteric liquid crystal structures with helicity inversion. *Opt. Commun.* 2009, 282, 903.

100. Relaix, S., Mitov, M. Polymer-stabilised cholesteric liquid crystals with a double helical handedness: Influence of an ultraviolet light absorber on the characteristics of the circularly polarised reflection band. *Liq. Cryst.* 2008, 35, 1037.

101. Guo, J., Cao, H., Wei, J., Zhang, D., Liu, F., Pan, G., Zhao, D., He, W., Yang, H. Polymer stabilized liquid crystal films reflecting both right- and left-circularly polarized light. *Appl. Phys. Lett.* 2008, 93, 201901.

102. Guo, J., Yang, H., Li, R., Ji, N., Dong, X., Wu, H., Wei, J. Effect of network concentration on the performance of polymer-stabilized cholesteric liquid crystals with a double-handed circularly polarized light reflection band. *J. Phys. Chem. C* 2009, 113, 16538.

103. Guo, J., Liu, F., Chen, F., Wei, J., Yang, H. Realisation of cholesteric liquid-crystalline materials reflecting both right- and left-circularly polarised light using the wash-out/refill technique. *Liq. Cryst.* 2010, 37, 171.

104. Guo, J., Wu, H., Chen, F., Zhang, L., He, W., Yang, H., Wei, J. Fabrication of multi-pitched photonic structure in cholesteric liquid crystals based on a polymer template with helical structure. *J. Mater. Chem.* 2010, 20, 4094.

105. Guo, J., Chen, F., Qu, Z., Yang, H., Wei, J. Electrothermal switching characteristics from a hydrogen-bonded polymer network structure in cholesteric liquid crystals with a double-handed circularly polarized light reflection band. *J. Phys. Chem. B* 2011, 115, 861.

106. Choi, S. S., Morris, S. M., Huck, W. T. S., Coles, H. J. Simultaneous red-green-blue reflection and wavelength tuning from an achiral liquid crystal and a polymer template. *Adv. Mater.* 2010, 22, 53.

107. Guo, J., Xing, H., Jin, O., Shi, Y., Wei, J. Electrically induced multicolored hyper-reflection and bistable switching from a polymer-dispersed cholesteric liquid crystal and a templated helical polymer. *Mol. Cryst. Liq. Cryst.* 2013, 582, 21.

108. McConney, M. E., Tondiglia, V. P., Hurtubise, J. M., Natarajan, L. V., White, T. J., Bunning, T. J. Thermally induced, multicolored hyper-reflective cholesteric liquid crystals. *Adv. Mater.* 2011, 23, 1453.

109. McConney, M. E., Duning, M. M., Natarajan, L. V., Voevodin, A. A., Tondiglia, V. P., White, T. J., Bunning, T. J. Tuning of the reflection properties of templated cholesteric liquid crystals using phase transitions. *Mol. Cryst. Liq. Cryst.* 2012, 559, 115.

110. McConney, M. E., White, T. J., Tondiglia, V. P., Natarajan, L. V., Yang, D.-k., Bunning, T. J. Dynamic high contrast reflective coloration from responsive polymer/cholesteric liquid crystal architectures. *Soft Matter* 2012, 8, 318.

111. Cazzell, S. A., McConney, M. E., Tondiglia, V. P., Natarajan, L. V., Bunning, T. J., White, T. J. The contribution of chirality and crosslinker concentration to reflection wavelength tuning in structurally chiral nematic gels. *J. Mater. Chem. C* 2014, 2, 132.

112. Jin, O., Xing, H., Jiewei, Guo, J. Photo-thermal modulation of cholesteric liquid crystals with a dual circularly polarized light reflection band. *Mol. Cryst. Liq. Cryst.* 2015, 608, 91.

113. McConney, M. E., Tondiglia, V. P., Hurtubise, J. M., White, T. J., Bunning, T. J. Photoinduced hyper-reflective cholesteric liquid crystals enabled via surface initiated photopolymerization. *Chem. Com.* 2011, 47, 505.

114. Zhao, Y., Zhang, L., He, Z., Chen, G., Wang, D., Zhang, H., Yang, H. Photoinduced polymer-stabilised chiral nematic liquid crystal films reflecting both right- and left-circularly polarised light. *Liq. Cryst.* 2015, 42, 1120.

115. Chen, L.-J., Lee, C.-R., Chu, C.-L. Surface passivation assisted lasing emission in the quantum dots doped cholesteric liquid crystal resonating cavity with polymer template. *Rsc Adv.* 2014, 4, 52804.

116. Sun, J., Yu, L., Wang, L., Li, C., Yang, Z., He, W., Zhang, C. et al. Optical intensity-driven reversible photonic bandgaps in self-organized helical superstructures with handedness inversion. *J. Mater. Chem. C* 2017, 5, 3678.

117. White, T. J., Cazzell, S. A., Freer, A. S., Yang, D.-K., Sukhomlinova, L., Su, L., Kosa, T., Taheri, B., Bunning, T. J. Widely tunable, photoinvertible cholesteric liquid crystals. *Adv. Mater.* 2011, 23, 1389.

118. Asshoff, S. J., Iamsaard, S., Bosco, A., Cornelissen, J. J. L. M., Feringa, B. L., Katsonis, N. Time-programmed helix inversion in phototunable liquid crystals. *Chem. Commun.* 2013, 49, 4256.

119. Ingle Jr, J. D., Crouch, S. R. *Spectrochemical Analysis*; Prentice-Hall: New Jersey, 1988, pp. 372–380.

120. Coates, D. Development and applications of cholesteric liquid crystals. *Liq. Cryst.* 2015, 42, 653.

121. Khandelwal, H., Loonen, R. C. G. M., Hensen, J. L. M., Debije, M. G., Schenning, A. P. H. J. Electrically switchable polymer stabilised broadband infrared reflectors and their potential as smart windows for energy saving in buildings. *Sci. Rep.* 2015, 5, 11773.

122. Zhang, W., Zhang, C., Chen, K., Wang, Z., Wang, M., Ding, H., Xiao, J. et al. Synthesis and characterisation of liquid crystalline anthraquinone dyes with excellent dichroism and solubility. *Liq. Cryst.* 2016, 43, 1307.

123. Zhang, W., Zhang, L., Liang, X., Zhou, L., Xiao, J., Yu, L., Li, F. et al. Unconventional high-performance laser protection system based on dichroic dye-doped cholesteric liquid crystals. *Sci. Rep.* 2017, 7, 42955.

124. Liu, Y., Yu, L., Jiang, Y., Xiong, W., Wang, Q., Sun, J., Yang, H., Zhao, X.-Z. Self-organized cholesteric liquid crystal polymer films with tunable photonic band gap as transparent and flexible back-reflectors for dye-sensitized solar cells. *Nano Energy* 2016, 26, 648.

2 Photochromic Chiral Liquid Crystals for Light Sensing

Ling Wang, Karla G. Gutierrez-Cuevas, and Quan Li

CONTENTS

2.1 Introduction 33
2.2 Photochromic Chiral Liquid Crystals... 35
    2.2.1 Photochromic Molecules 35
    2.2.2 Chiral Liquid Crystalline Phases... 36
    2.2.3 Fabrication of Photochromic Chiral Liquid Crystals 38
2.3 Photochromism in 1D Chiral Liquid Crystalline Superstructures................ 38
    2.3.1 Photochromic CLCs Containing Chiral Azobenzene Dopants.......... 39
    2.3.2 Photochromic CLCs Containing Chiral Diarylethene Dopants......... 42
    2.3.3 Photochromic CLCs Containing Chiral Spirooxazine and
            Overcrowded Alkenes.. 44
2.4 Photochromism in 3D Chiral Liquid Crystalline Superstructures................ 47
2.5 Photochromic Liquid Crystals for Potential Infrared Sensors 51
2.6 Summary and Perspectives... 55
Acknowledgments.. 55
References... 56

2.1 INTRODUCTION

The sun, which is fundamental source of light for the existence of life on our planet, emits electromagnetic radiation in a broad spectral range. Despite the undeniable importance, such as cutaneous exposure to some ultraviolet radiation for vitamin D production and 49% of the heating of earth from infrared irradiation of the sun, there is convincing evidence that the human skin and eyes are adversely affected in many different ways by their exposures to natural sunlight and artificial light sources such as lasers.[1-3] Excessive exposures to these light radiations are also found to result in the premature aging, deterioration, or expiration of perishable food and high active chemicals or materials. To increase people's awareness of these potential hazards, many monitoring systems have been developed for environmental light detection and laboratory hazardous measurements; but these expensive and complicated equipments are generally not suitable for use by the public.[4-7] Moreover, since irradiation levels are often dependent on the local environmental conditions such as air pollution index and the intensity of ambient light, it is clear that not only many monitoring sites are needed but also the monitoring must be carried out continuously over a long period of time.

Photochromic sensor is emerging as a candidate for light-weight, inexpensive, and easy-to-use light detector that can be set up in every corner of the world. The originality of photochromic sensor is its ability to manifest a discernible amount of visible colors under sunlight without the aid of any filter, photodiode, or any kind of electronic amplifying circuit.[8–10] To be specific, photochromic compounds are typically colorless in their initial states but turn colored upon light irradiation, which helps the human eye to directly "see" and even "feel" their intensities. Importantly, this kind of photochromic detectors can be easily incorporated in any portable daily items, such as watches, pins, buttons, stickers, sportswear, telephone cards, and credit cards. Furthermore, the incorporation of photochromic sensors into the tags or packages would be also helpful to alert consumers about the conditions of the perishable food and high active chemicals or materials. However, the present photochromic sensors exhibit the monotonous color change from colorless state upon response to the light irradiations. Therefore, there are tremendous incentives to develop new photochromic sensors displaying diverse color changes in a multitude of variable red, green, and blue colors.

Liquid crystals (LCs) have become quintessential materials in our daily life. The creation of a multibillion dollar LC display (LCD) industry for information displays demonstrates how great economic prosperity can evolve from a pure curiosity-driven academic research endeavor. Display devices using LCs as the active components like mobile phones, computer and television screens, projectors, etc. have enhanced the quality and comfort of our life and have drastically revolutionized the way we present the information.[11–14] Such technologies are based on the reorientation of liquid crystalline materials in response to the applied electric field, which results in a change in its observed optical properties. It should be noted that most present commercial displays are based on the LCs exhibiting a nematic phase, which is the simplest phase in the field of LCs.

Introducing chirality into the nematic liquid crystalline phase often leads to novel self-organized helical superstructures, that is, cholesteric LCs (CLCs).[15,16] Fundamentally, one of the key features that these CLC materials possess is the existence of a photonic bandgap within the visible regions. Such photonic band structures have attracted considerable attention in recent years because of their ability to exhibit tunable selective reflection of visible light under various external stimuli, such as light, heat, pressure, electric field, magnetic field, and chemical reactions.[17–23] This makes it possible to use CLCs as various smart sensors where signal changes upon response to different external perturbations can be easily observed with the naked eye.[24–27] Interestingly, CLCs that reflect circular polarized light at a given wavelength as a result of the self-organized helical superstructures have been utilized for constructing chemical sensors that do not require batteries. The pitch of the helix in these CLCs determines the wavelength of photonic reflection and can be modified in response to the chemical atmosphere such as gas, humidity, pH, and CO_2, resulting in a color change.[28–30] These appealing smart CLC materials would find wide applications in the built environment, packaging industry, health care, etc. It is envisioned that combining photochromic behavior of light-driven molecular switches or motors with photonic reflection of various chiral LCs such as CLCs and blue phases (BPs) would provide significant impetus for developing novel light sensors that help to increase the public awareness regarding health and environmental issues especially related to the invisible harmful or hazardous radiations.

The subject of this chapter is confined to the survey of photochromic chiral LCs for potential light sensing applications. The introduction of photochromic chiral switches or motors in different LC hosts enables the dynamic circularly polarized visible light reflections under light irradiation, and photochromic sensors based on these liquid crystalline materials would be very helpful to alert consumers about the conditions of perishable products and increase public awareness of hazardous irradiations from natural sunlight and artificial light sources. This chapter is organized in the following manner. The first section introduces the photochromic chiral LCs including photochromic molecules and chiral liquid crystalline phases. Second, we examine the photochromism in one-dimensional (1D) liquid crystalline superstructures, that is, CLCs followed by the discussion of photochromism in three-dimensional (3D) liquid crystalline superstructures, that is, BPs and liquid crystalline microdroplets and microshells. Before we conclude this chapter with a perspective for the future scope and challenges for these emerging photochromic chiral LCs in potential sensing applications, the recent endeavor in developing photochromic LCs for potential infrared sensors is presented.

2.2 PHOTOCHROMIC CHIRAL LIQUID CRYSTALS

2.2.1 PHOTOCHROMIC MOLECULES

Photochromism is the term used for reversible photoinduced transformation of a molecule between different isomers whose absorption spectra are distinguishably different.[31] During the reversible photoisomerization of photochromic molecules, some physical properties, such as absorption spectra and geometrical structure, can be tuned by light irradiations. Generally, two stable isomers are involved, and they undergo the interconversion under ultraviolet (UV), visible, or infrared irradiation. By the modulation of irradiation light wavelength, photochemical reaction occurs between open and close, *trans-* and *cis-*isomerization, etc. The molecular shape/geometry/conformation change of photochromic dopants upon the light irradiation constitutes the basis for dynamic phototuning of the properties of photochromic chiral LCs.[32] Typical photochromic molecules including azobenzene, dithienylethene, spirooxazine, and their photoisomerization processes are shown in Scheme 2.1.

Azobenzene derivatives are a well-known family of photochromic compounds that can experience *trans–cis* isomerization upon UV irradiation.[33] The *cis* isomer can be driven back to its *trans* form either by visible light or heat. The rod-like structure of *trans* form can stabilize calamitic LCs, while the *cis* form is bent and usually decreases the order parameter of LC phases. Owing to the dramatic shape change between the *trans* and *cis* isomers, azobenzenes were intensively investigated as mesogens or dopants in photochromic chiral LCs.

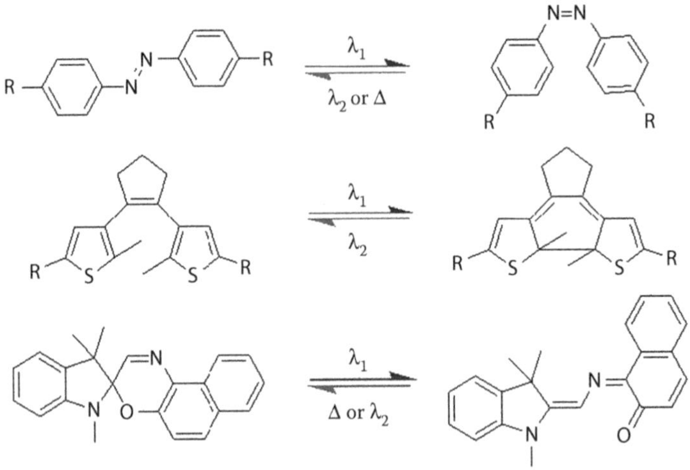

SCHEME 2.1 Molecular structures and photoisomerization of photochromic molecules including azobenzene, dithienylcyclopentene, and spirooxazine.

Diarylethenes are also a fascinating class of photochromic molecules due to their superior thermal stability and excellent fatigue resistance.[34,35] Upon UV irradiation, they can transform from the colorless open-ring form to the colored closed-ring form. The reverse process is thermally stable and occurs only by visible light irradiation. Compared with extensively studied azobenzene-based molecules for photochromic chiral LCs, there are a few reports on thermally stable chiral diarylethenes.

The other family of photochromic molecules is spiropyrans and spirooxazines.[36–38] Upon UV irradiation, photochemical cleavage of the C–O bond results in the transformation from the initial colorless state to the colored merocyanine. However, when the UV source is removed, they usually can thermally relax back to their colorless initial state quickly. Spirooxazines are particularly interesting due to their unique properties such as excellent photofatigue resistance, strong photocoloration, and fast thermal relaxation.[39] Since the physical and chemical properties of the two forms are dramatically different, the thermally reversible photochromic switching has been the basis for the intelligent materials with applications in various photonic devices including photochromic sensors. However, to date, there have been only very few reports on chiral spirooxazines.

2.2.2 Chiral Liquid Crystalline Phases

There are many types of chiral liquid crystalline phases such as cholesteric phases, chiral smectic phases, and BPs; however, in this chapter, we restrict our focus on cholesteric phases and BPs for light sensing applications. It should be noted that cholesteric phases are considered to be self-assembled 1D photonic crystals whereas the BPs are regarded as self-assembled 3D photonic crystals.

Cholesteric phase, also known as chiral nematic phase, is the LC phase with chirality at the molecular level and helical arrangement of layers at the macroscale level.[40] CLCs are often developed by doping chiral molecular switches into an achiral nematic host. The ability of the chiral dopant to twist the achiral LCs is defined as helical twisting power (HTP, β) and can be expressed as $\beta = 1/(pc)$, where c is the chiral dopant concentration and p is the pitch of helical superstructure. The helical superstructure of CLC is characterized by helical pitch and handedness. The pitch (p) is the distance across which the director rotates a full 360°. Handedness describes the direction in which the molecular orientation rotates along the helical axis and it can be expressed as sign ($-$) and ($+$), which represent left and right handedness, respectively. The most important optical property of CLC is its selective reflection of light. When unpolarized light propagates through a CLC medium, only the circularly polarized light (CPL) with the same handedness as its helix is reflected, that is, left-handed CLCs only reflect left-handed CPL and right-handed CLCs only reflect right-handed CPL. The reflection wavelength of a cholesteric phase can be determined by $\lambda = np$, where n is the average refraction index of LC material. The helical superstructure of CLCs with the pitch length in the range of hundreds of nanometers can reflect light in the visible range and the reflection color tuning has been utilized for tunable color reflectors and filters, reflection displays, and biomedical applications.[41] With photochromic molecules or moieties in CLCs, it is possible to tune the pitch length of CLCs under the light irradiation (Figure 2.1), and the resultant photoinduced color changes provide numerous opportunities for applications in photochromic optical sensors.

BPs are among the most fascinating soft photonic nanostructures in the area of chiral liquid crystalline phases. They are known to self-organize into the 3D frustrated nanostructures, which originate from the competition between the packing topology and chiral forces. BPs generally exist within the temperature range between the isotropic phase and cholesteric phase.[42–45] The cubic structures are characterized by the double-twist cylinders (DTCs) which are stabilized by defects or disclinations. There are three types of BPs reported, namely BP III, BP II, and

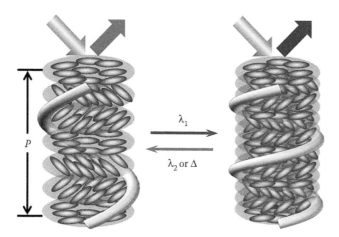

FIGURE 2.1 Schematic mechanism of the reflection wavelength of photochromic chiral LCs reversibly and dynamically tuned by light irradiations.

BP I, which are observed during cooling from the isotropic to the cholesteric phase in the order of decreasing temperature. BP I and BP II have body-centered cubic and simple cubic structures, respectively, while BP III is an amorphous network of disclination lines.[46–51] It is noteworthy that BP I and BP II are periodic in three dimensions and the periodicity, that is, lattice constant is typically on the order of the wavelength of visible light. Thus, these nanostructures enable BPs to reflect light in the visible region and as mentioned before BPs have been recently recognized as soft 3D photonic crystals.[52,53] The reflection wavelength of BPs containing photochromic molecules can be tuned by light-driven changes in their lattice constants. Compared with simple 1D helical superstructures of CLCs, the 3D cubic structures of BPs offer several advantages on device performance over cholesteric materials such as fast response speed, high contrast ratio, and no surface treatment.

2.2.3 Fabrication of Photochromic Chiral Liquid Crystals

Introducing photochromism and chirality into LCs enables us to develop the photochromic chiral LCs with novel and enhanced properties for practical applications.[54–58] In general, there are two main strategies to fabricate photochromic chiral LCs. The first and simplest one is based on single-component chiral mesogens with linked photochromic moiety, which can exhibit the liquid crystalline phase at a certain temperature range. The photochromic chiral LCs produced by this method consist of pure materials and thus may exhibit several advantages such as good uniformity and enhanced stability. However, it usually suffers from the high synthetic cost of photochromic chiral mesogens, the requirement of above-ambient-temperature processing conditions, and physical properties that are not suitable for devices. The second strategy utilizes the host–guest system, in which LC phases of calamitic (rod-like) mesogens are generally used as the host and photochromic chiral molecules are doped as guests.[59–63] In the doping system, it is relatively easy to modulate the LC photonic properties by adjusting the doping concentration of the guest chiral molecules. Furthermore, the photochromic chiral dopant can be synthesized separately from the LC host and the use of a small amount of chiral switch means much lower cost. It is noteworthy that this strategy is strongly dependent on the development of the powerful photochromic molecular switch with high chiral induction capability, fast response photochromic moieties, and good compatibility with LC host.

2.3 PHOTOCHROMISM IN 1D CHIRAL LIQUID CRYSTALLINE SUPERSTRUCTURES

To date, extensive efforts have been devoted to developing various photochromic chiral molecules which are widely applied for phototuning the dynamic reflection colors of CLCs, that is, photochromism in 1D liquid crystalline superstructures. Photoisomerization of the photochromic chiral switches results in changes in the molecular conformation and HTP, which would lead to a change in the pitch of the CLCs and their reflection color variations under the light irradiations. One of critical challenges in this endeavor is the design and synthesis of photoresponsive chiral molecular switches with high HTP values that display large differences in HTP in their different isomeric states.

2.3.1 Photochromic CLCs Containing Chiral Azobenzene Dopants

In 1971, Sackmann reported a light-induced reversible color change in nonhalide-based cholesteryl mixture doped with azobenzene.[64] The photonic bandgap reflection of the CLC mixture blueshifted from 610 to 560 nm upon irradiation with 420-nm light. Although the tuning range is quite narrow, this work is the first demonstration of reversible phototuning of cholesteric reflection colors, providing a pathway to the photoresponsive CLCs. With the continuous efforts especially in the past decade, many chiral azobenzene dopants have been developed and used for phototuning the reflection colors of CLCs with large tuning range, very high HTPs, and low doping concentrations.

Tamaoki et al. reported a series of planar chiral dopants which were employed in commercially available nematic LCs to achieve phototunable reflection colors.[65] These compounds were designed based on an azobenzenophane compound having conformational restriction on the free rotation of naphthalene moiety to impose an element of planar chirality. Due to the good solubility, moderately high HTPs, and large changes in HTPs during the photoisomerization of dopant, a fast photon mode reversible color control in induced CLCs was achieved. Moreover, systematic investigation indicated that the number of chiral groups introduced in a photochromic molecule strongly influenced the initial HTP.[66,67] Azobenzene compounds substituted with two chiral groups at both ends of the azo core were found to possess a much higher initial HTP as compared with the azobenzene compounds substituted with a single chiral group at one end of the azo core. However, the photochemical change in HTP of the disubstituted azobenzene compounds was smaller than that of monosubstituted ones. Therefore, Kurihara et al. synthesized a series of isosorbide- or isomannide-based chiral photochromic compounds with multiple azobenzene groups, and found that chiral azobenzene **1** as shown in Figure 2.2a exhibited high HTPs as well as large HTP variations upon light irradiation.[68] A large tuning range of ~500 nm was reported in the CLCs doped with chiral azobenzene **1** and another nonphotoresponsive chiral compound as dopants combined with LC host E44. Interestingly, the color could also be adjusted by varying the light intensity with a gray photomask, as seen in Figure 2.2b and c. The resolution of the color patterning was estimated to be 70–100 μm by patterning experiments with the use of a photomask.

Binaphthyl derivatives with axial chirality are known as powerful helicity inducers. The unique combinations of binaphthyl with azobenzene units were found to generate a family of axially chiral azobenzene compounds with very high HTPs and significantly broaden the area of reflection color tuning using a single chiral photoresponsive dopant.[69] We reported four reversible photoswitchable axially chiral azo dopants **2a–2d** with high HTPs as shown in Figure 2.3a.[70] These light-driven chiral switches were also suitable for dopants in achiral nematic host for applications in novel optical addressed displays, that is, photodisplay. For example, an image was created on the display cell filled with chiral switch **2a**-based CLC using UV light

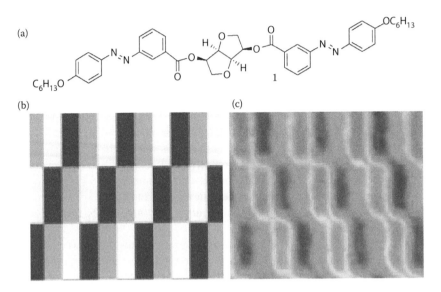

FIGURE 2.2 (a) Molecular structure of chiral azobenzene **1**. (b) Gray mask. (c) Resultant patterning of photochromic chiral LCs obtained by UV irradiation for 10 s through the gray mask at 25°C. Note: For colorful patterning, please check the original figure in the reference. (Yoshioka, T et al.: *Adv. Mater.* 1226–1229. 2005. Copyright Wiley-VCH Verlag GmbH & Co. KGaA. Reproduced with permission.)

with a negative photomask and exposed to UV light for 20 min. Depending on the optical density distribution of the mask, certain areas were exposed with different intensities of light, resulting in an image composed of a variety of colors due to the various shifts in pitch length. Figure 2.3b–d shows the photograph of an original image, the negative mask, and the resulting image on the display cell, respectively. The light-driven chiral molecular switches in LC media were sufficiently responsive to an addressing light source so that a high-resolution image with gray scale could be imaged in a few seconds of irradiation time.

Photochromic compound **3** as shown in Figure 2.4a exhibits improved properties with higher HTPs than molecular switches **2**.[71] The switch was found to impart its chirality to a commercially available nematic LC host, at low doping levels, to form a self-organized, optically tunable helical superstructure capable of fast and reversible phototuning of the structural reflection across entire visible region. For example, a mixture of 6.5 wt% **3** in nematic LC E7 was capillary filled into a 5-μm thick glass cell with a polyimide planar alignment layer and the cell was painted black on one side. The reflection wavelength of the cell could be tuned starting from UV region across the entire visible region to near-infrared (NIR) region upon UV irradiation at 365 nm (5.0 mW/cm^2) within ~50 s, whereas its reversible process starting from NIR region across the entire visible region to UV region was achieved by visible light at 520 nm (1.5 mW/cm^2) or dark thermal relaxation. The reflection colors across the entire visible region were uniform and brilliant as shown in Figure 2.4b and c. The reversible phototuning process was repeated many times without degradation. It is

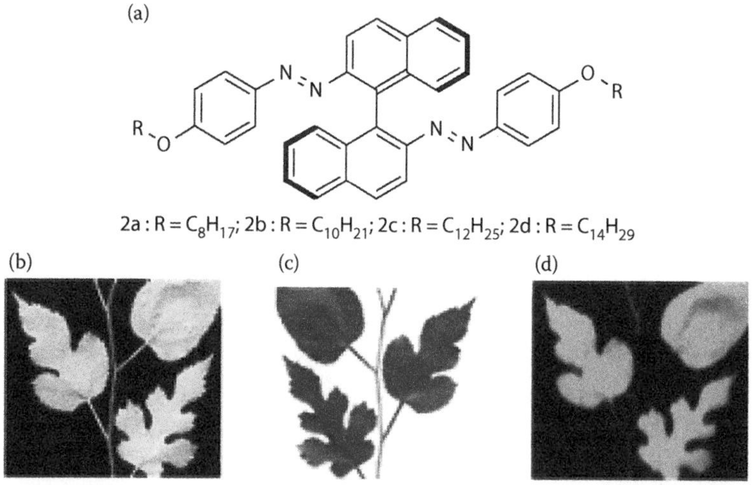

FIGURE 2.3 (a) Molecular structures of chiral azobenzene **2a–2d** with axial chirality. Illustration of an optically addressed image with negative photomask. (b) Regular photograph of the original digital image. (c) Negative photomask made of polyethylene terephthalate (PET). (d) Image optically written on the display cell. (Reprinted with permission from Li, Q et al. Reversible photoswitchable axially chiral dopants with high helical twisting power. *J. Am. Chem. Soc.* 129, 12908–12909. Copyright 2007 American Chemical Society.)

worth noting that the reversible phototuning process across the entire visible region can be achieved in seconds with the increase of light intensity, and the photochromic sensors using this kind of materials might help the human eye to "see" the light irradiations and "feel" their intensities at the same time. Furthermore, this chiral switch **3** was used in a photo-addressed colored LCD driven by light and hidden as well as fixed by application of an electric field from thermal degradation. As illustrated in Figure 2.4d, the reflective image can be hidden in focal conic texture by applying a 30 V pulse and revealed by applying a 60 V pulse. Moreover, by applying a 30 V pulse to an optically written image so as to make the UV-irradiated region going to the focal conic texture and the UV-unirradiated region going to the planar texture, an optically written image can be stored indefinitely because the planar and focal conic textures are stable even though the light-driven switch relaxes to the unirradiated state.

Moreover, molecular switches **4a** and **4b** as shown in Figure 2.5a exhibit unusually high HTP values in their ground state (*trans–trans*) as well as a considerable reduction in HTP upon exposure to UV light (365 nm), which drives the molecules toward the *cis–cis* configuration.[72] The compound **4a** exhibited a very high HTP of 301 μm^{-1} at the initial state, which significantly decreased to 106 μm^{-1} at the photostationary state (PSS) upon UV irradiation. When chiral switch **4a** was doped into a commercial achiral LC host E7 at low concentrations, thin films which exhibited a fast and dynamically reversible response to light were enabled and reflection color tuning from blue to green and further to red upon UV irradiation and vice versa upon visible light was obtained. Very interestingly, it was found that the HTP of dopant **4a** was sensitive to the wavelength of exposure during the reverse process as the equilibrium ratio among

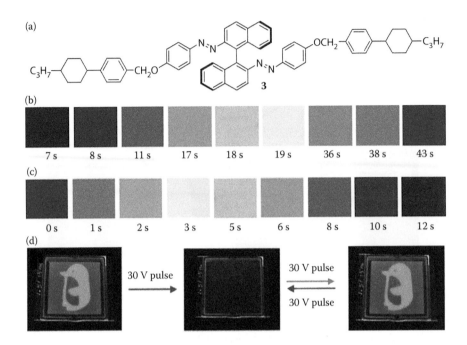

FIGURE 2.4 (a) Molecular structure of chiral azobenzene **3** with axial chirality. Reflection optical images of 6.5 wt% chiral switch **3** in commercially available achiral LC host E7 in 5-μm thick planar cell; (b) upon UV light at 365 nm (5.0 mW/cm²) with different time exposure; (c) reversible back across the entire visible spectrum upon visible light at 520 nm (1.5 mW/cm²) with different time exposure. The optical images were taken from a polarized reflective mode microscope. (d) Images of 5-μm thick homeotropic alignment cell with 4 wt% chiral switch **3** in E7. The image was recorded in a planar state through a photomask by a UV light (left). The image was hidden by a low-voltage pulse in a focal conic state (middle), which was reappeared by a high-voltage pulse (right). Note: For the colorful images, please check the original figures in the reference. (Ma, J et al. 2010. Light-driven nanoscale chiral molecular switch: Reversible dynamic full range color phototuning. *Chem. Commun.* 46: 3463–3465. Reproduced by permission of The Royal Society of Chemistry.)

trans–trans, trans–cis, cis–cis isomers was varied. As illustrated in Figure 2.5b–d, three stable primary colors red, green, and blue were realized at the PSS upon irradiation with 440, 450, and 550-nm visible light, respectively. The wavelength-selective reflection in this type of photochromic chiral LCs was expected to help the human eye to "see" the light irradiations and "feel" their wavelengths at the same time.

2.3.2 PHOTOCHROMIC CLCs CONTAINING CHIRAL DIARYLETHENE DOPANTS

Compared with photochromic chiral azobenzene switches, diarylethenes possess the advantage of thermal irreversibility in both the open- and closed-ring isomeric states. However, only a few derivatives of diarylethenes have been reported as chiral dopants with applications as liquid crystalline phase switches.[73–77] Most of reported chiral diarylethenes exhibited low to moderate HTPs and are not suitable for reflection

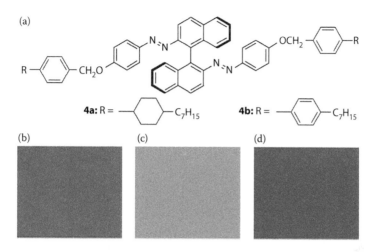

FIGURE 2.5   (a) Molecular structures of chiral azobenzene **4a** and **4b** with axial chirality. (b), (c), (d): Photostationary reflection images upon visible light irradiation at 440, 450, and 550 nm from UV-irradiated state, respectively. The reflection optical images were from 6.0 wt% chiral switch **4a** in LC host E7 in a 5-μm thick planar cell taken from a polarized reflective mode microscope. Note: For the colorful images, please check the original figures in the reference. (Li, Q et al.: *Adv. Mater.* 5069–5073. 2011. Copyright Wiley-VCH Verlag GmbH & Co. KGaA. Reproduced with permission.)

wavelength tuning. These relatively low HTPs also give rise to the requirement of higher doping concentrations, which often leads to phase separation, coloration, and physical property changes of LC host. Therefore, it would be of great practical interest to develop chiral diarylethenes with high HTPs as well as large HTP variations upon light irradiation. It was not until recently that we reported some chiral diarylethene switches for wide-range reflection color tuning.[76,77] For example, chiral switch (S,S)-**5** as shown in Figure 2.6a was found to possess remarkable changes in HTP during photoisomerization in addition to the very high HTP at the initial state. When doping 0.4 wt% of compound (S,S)-**5** into the LC host, the liquid crystalline phase could be transformed from nematic to cholesteric due to the dramatic increase in HTP upon UV irradiation. A higher doping concentration of 7.7 wt% was used for the phototuning of reflection colors of CLC mixtures. The reflection central wavelength of this mixture was around 630 nm at the initial state. Upon UV irradiation, its reflection wavelength was tuned to 530 nm within 10 s and further reached a PSS in 25 s with a reflection central wavelength at 440 nm. This photoinduced state was thermally stable and could be photochemically switched back to a nearly initial state by visible light irradiation at 550 nm within 2 min. When this cell was stored in the dark at any irradiated state, no observable change was found in either reflection color or reflection wavelength, even after 1 week, which results from the excellent thermal stability of compound (S,S)-**5**. Furthermore, three primary red, green, and blue colors can be observed simultaneously in a single thin film based on different UV irradiation times facilitated by masking at different areas: red, no irradiation; green, irradiated for 10 s; blue, irradiated for 25 s (Figure 2.6b). After driving the background color to

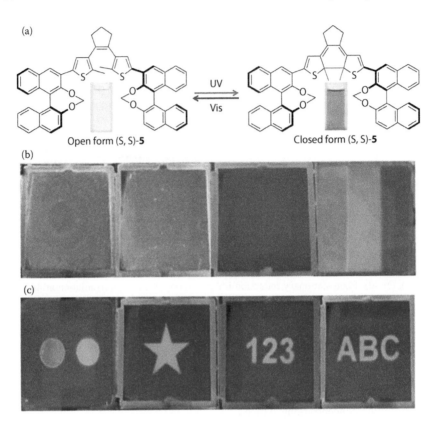

FIGURE 2.6 (a) Photoisomerization of chiral diarylethene dopant (*S,S*)-**5**. (b) Real cell images of an 8-μm thick planar cell (2.1 cm × 2.5 cm) filled with 7.7 wt% of (*S,S*)-**5** in E7. Reflection color change: red, green, blue, red–green–blue colors in one cell, from left to right. (c) Optically addressed images (green images on blue background). Note: For the colorful images, please check the original figures in the reference. (Reprinted with permission from Li, Y. et al. Reversible light-directed red, green, and blue reflection with thermal stability enabled by a self-organized helical superstructure. *J. Am. Chem. Soc.* 134, 9573–9576. Copyright 2012 American Chemical Society.)

blue by UV irradiation, the red and green reflection colors can be recorded through visible light irradiation for different times (Figure 2.6c). Furthermore, the optically addressed images can be erased by light irradiation when desired, and the cell is rewritable for many times due to the excellent fatigue resistance.

2.3.3 PHOTOCHROMIC CLCs CONTAINING CHIRAL SPIROOXAZINE AND OVERCROWDED ALKENES

Spirooxazines are an interesting family of photochromic materials due to their unique properties such as excellent photofatigue resistance, strong photocoloration, and fast thermal relaxation.[78] The fatigue resistance of spirooxazines has led to their successful use in various applications including commercialized eyewear. This kind of photochromic materials showing such intense photocoloration and fast thermal bleaching performance could be highly promising materials for photochromic sensors with high sensitivity. Since the spirocarbon of a spirooxazine molecule has potential as a chiral center, spirooxazines could be used as chiroptical molecular switches.[79] However, spirooxazines usually exist as racemates. Even if enantiomers are separated, each enantiomer is racemized by thermal and optical interconversions. Therefore, if spirooxazines are to be utilized as chiroptical molecules in achiral nematic LC systems, modification of the spirooxazine with a chiral group is required. Accordingly, we synthesized series of axially chiral spirooxazines as shown in Scheme 2.2 which exhibit high HTP.[80] Photochromic CLCs were fabricated by doping these photochromic molecules 6a–6d into commercially available nematic LC hosts. Among these chiral dopants, the compound 6a with bridged binaphthalene group showed the largest HTP value in E7. Interestingly, for compounds 6a and 6c, the HTP increases upon irradiation with UV light, while the HTP of other compounds decreases under the same condition. This observation has been attributed to more compatibility of the rod-like merocyanine form for compounds 6a and 6c compared with the other compounds. However, there is a long way to achieve the molecular switches that exhibit high HTPs as well as large HTP variations upon light irradiation for applications of photochromic chiral LC sensors.

In addition to chiral azobenzene, diarylethene, and spirooxazine, chiral overcrowded alkenes have been also intensively studied. These types of molecules are also found capable of tuning the reflection colors in CLCs. Take light-driven chiral motor

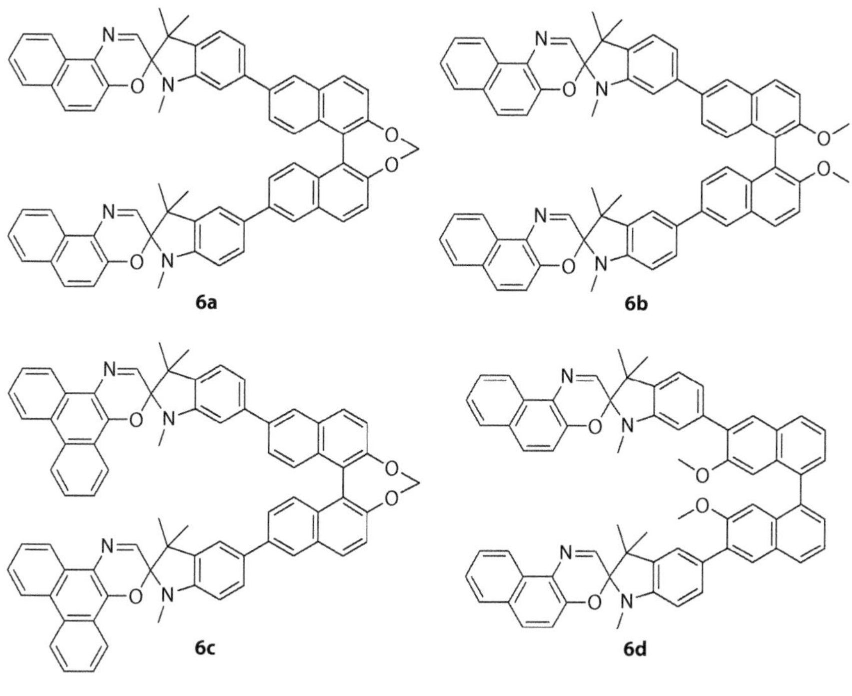

SCHEME 2.2 Molecular structures of chiral spirooxazine molecular switches 6a–6d.

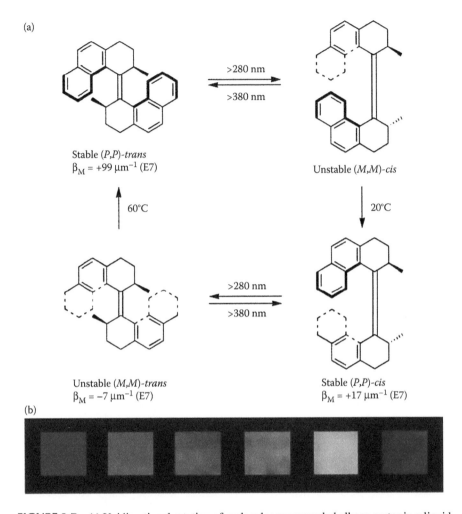

(a)

Stable (P,P)-*trans*
β_M = +99 µm⁻¹ (E7)

Unstable (M,M)-*cis*

Unstable (M,M)-*trans*
β_M = −7 µm⁻¹ (E7)

Stable (P,P)-*cis*
β_M = +17 µm⁻¹ (E7)

(b)

FIGURE 2.7   (a) Unidirectional rotation of molecular overcrowded alkene motor in a liquid crystalline host, and associated HTPs. (b) Optical images of overcrowded alkene-doped LC phase (6.16 wt% in E7) in time, starting from pure (P, P)-*trans* overcrowded alkene upon irradiation with >280-nm light at room temperature, as taken from actual photographs of the sample. The images shown from left to right correspond to 0, 10, 20, 30, 40, and 80 s of irradiation time, respectively. Note: For the colorful images, please check the original figures in the reference. (van Delden, R. A. et al. Supramolecular chemistry and self-assembly special feature: Unidirectional rotary motion in a liquid crystalline environment: Color tuning by a molecular motor. *Proc. Natl. Acad. Sci.* USA 99: 4945–4949, Copyright 2002 National Academy of Sciences, U.S.A.)

in Figure 2.7 as an example.[81] Its initial HTP at (P, P)-*trans* form in nematic E7 is +99 µm⁻¹, but generation of a cholesteric helix with an opposite sign of similar pitch is impossible, as the (M, M)-*trans* form possesses a minor negative HTP (β_M = +17 µm⁻¹, E7). As a result of the high HTP at (P, P)-*trans* form, colored LC films were easily generated using this dopant. Photochemical and thermal isomerizations of the motor lead to irreversible color change in the LC film as shown in Figure 2.7b (bottom).[82] Another iteration of this class of molecules was also demonstrated to have a broader tuning range covering the entire visible region with reversibility.[83] However, the drawback of this system is that the reverse process is relying on long-time thermal relaxation.[84]

2.4 PHOTOCHROMISM IN 3D CHIRAL LIQUID CRYSTALLINE SUPERSTRUCTURES

Although 3D liquid crystalline superstructures are not necessarily required for some basic optical sensing, which is possible with simpler 1D cholesteric superstructures as aforementioned, existence of multiple photonic bandgap regions along different directions in the 3D superstructures might add an interesting capability of performing simultaneous optical sensing at several wavelengths along different directions.[85–87] Interestingly, laser emissions in three orthogonal directions have been detected simultaneously in the 3D photonic BP II (Figure 2.8).[52] When one of platelet BP domains in a LC thin cell was excited by the pump laser beam, laser emissions in the orthogonal (100), (010), and (001) directions can be observed at the same time, which is

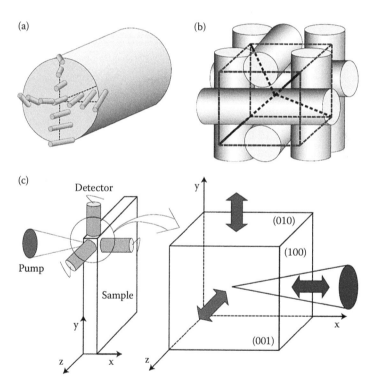

FIGURE 2.8   (a) Schematic illustration of the double-twist structure and (b) the resultant simple cubic BP II. (c) Regions of the sample from which the laser emissions can be detected in three orthogonal directions. (Reprinted by permission from Nature Publishing Group. *Nat. Mater.*, Cao, W. et al. Lasing in a three-dimensional photonic crystal of the liquid crystal blue phase II. 1: 111–113, copyright 2002.)

actually a sign of the distributed feedback (DFB) in three dimensions. Light-driven self-organized BP 3D cubic nanostructures that display distinctive photonic reflection in three orthogonal directions have received increasing attentions especially in recent years, but the reflection wavelength tuning is usually quite narrow.[88–90] The ability to dynamically tune the photonic bandgap in cubic BPs across a wide wavelength range is highly desirable, but it has not been realized due to various obstacles including the instability and irreversibility of BPs under the light irradiations.

Very recently, a breakthrough of full visible range reflection phototuning in BPs was achieved by using an axially photochromic chiral azobenzene dopant **7** as shown in Figure 2.9a.[91] The initial phase of the doped BPs was the BP II with (100) lattice confirmed by the Kossel diagram as shown in Figure 2.9d (0 s) with a blue reflection

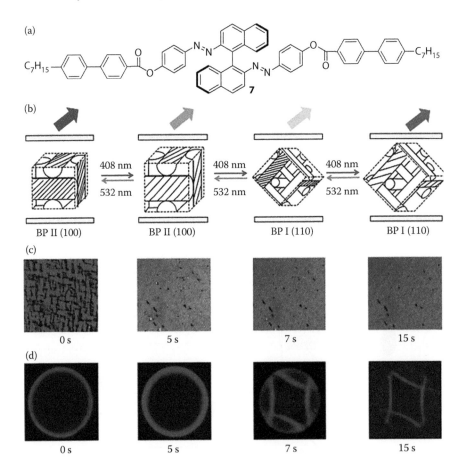

FIGURE 2.9 (a) Molecular structure of chiral azobenzene **7** with axial chirality. (b) Schematic illustration of BP cubic nanostructures. (c) Reflection optical images of BPs upon 408-nm light irradiation. (d) The corresponding optical Kossel diffraction diagrams. Note: For the colorful images, please check the original figures in the reference. (Lin, T. H. et al.: *Adv. Mater.* 5050–5054. 2013. Copyright Wiley-VCH Verlag GmbH & Co. KGaA. Reproduced with permission.)

color as shown in Figure 2.9c (0 s). Upon 408-nm light irradiation, the photoisomerization of the azobenzene dopant enlarged the size of the BP lattice and shifted the corresponding reflection toward a longer wavelength. Increasing the irradiation time resulted in occurrence of the phase transition from BP II to BP I, which might be attributed to the chirality decrease with dopant isomerization.[92] After the photoinduced phase transition took place, the lattice direction changed from (100) of BP II to (110) of BP I with the corresponding change of reflection wavelength from 520 to 575 nm. Subsequently, a redshift in reflection upon further irradiation at 408 nm was observed in BP I. If the irradiation was stopped at any time, the wavelength of the selective reflection does not change, except by very slow thermal relaxation effect of the dopant toward its initial *trans* form. The rapid reverse process can occur photochemically with a 532-nm light irradiation. Figure 2.9c shows BP images with various durations of pumping observed under a crossed polarized optical microscope at reflection mode. Upon 408-nm light irradiation, the reflection color of the BP thin film changed from blue (BP II) to red (BP I) within 15 s. The BP thin films were also indexed using a Kossel diagram as shown in Figure 2.9d. Both blue and green images in Figure 2.9c corresponded to (100) lattices of BP II, while both yellow-green and red images represented (110) lattices of BP I. Figure 2.9d also exhibited the variation of the Kossel diagram with different exposure time at 408 nm. As the irradiation time increased from 0 to 15 s, the circle patterns of the (100) lattice of the BP II enlarged, switching to diamond-shaped pattern corresponding to (110) lattice of BP I, and gradually shrinking. It is noteworthy that increasing the dopant concentration shifted the initial BP reflection to the UV region, which was tuned across the entire visible region with broad range upon 408-nm light irradiation, the reflection bands of BP II redshifted from the UV to 520 nm. Once the BP II transformed into the BP I, the corresponding reflection bands discontinuously jumped to 560 nm, which was further redshifted from 560 to 710 nm within BP I state.

Self-organized 3D superstructures with spherical configurations have recently received considerable attentions in the forefront of multitudinous research areas, such as optical sensors or manipulations, soft photonics, and microfluidics.[93] Liquid crystalline microdroplets and microshells, that is, confining LCs in micrometer- or nanometer-sized superstructures, represent an elegant example of the topic. Recent advances of microfluidic techniques have enabled the fast, simple, and convenient fabrication of stable 3D monodisperse microdroplets with controllable symmetry, shapes, and sizes.[94] With the help of microfluidic techniques, the size-polydispersity obstacles during the processes of producing the microdroplets have been successfully overcome, which opens infinite possibilities for 3D optical sensors.[95,96] Recently, we demonstrated the photonic cross-communication between CLC monodisperse microdroplets, where their self-assembly was precisely manipulated by using the microfluidic method.[97] Only a central reflection spot was observed in the isolated single microdroplet due to the absence of any lateral communication (Figure 2.10a). In the groups of two or more microdroplets, it was surprising that the reflection-related interactions were found between neighboring microdroplets (Figure 2.10b–e). When the microdroplets are densely packed into the hexagonally symmetric monolayer, the circular "lit firecracker" patterns were formed because of the photonic cross-communication in the neighbor microdroplets (Figure 2.10f).

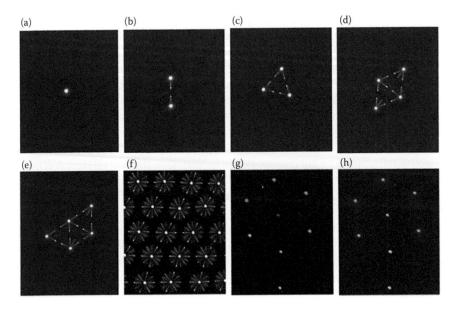

FIGURE 2.10 (a–f): Polarizing optical microscopy images and photonic cross-communication of linear-, triangular-, and diamond-shaped patterns in the cholesteric microdroplet arrays. (g–h): The "flower-opening" patterns of microdroplets with light-driven iridescent colors. (Fan, J. et al.: *Angew. Chem. Int. Ed. Engl.* 2188–2192. 2015. Copyright Wiley-VCH Verlag GmbH & Co. KGaA. Reproduced with permission.)

Moreover, the intensity of photonic cross-interaction was found to depend strongly on the distance between the microdroplets, and the photonic cross-communication completely disappeared when the distance was five times larger than the diameter of microdroplets. It is worth mentioning that omnidirectional red–green–blue reflections in the microdroplets have been achieved by using a novel thermally stable photochromic chiral molecular switch. We could also readily modulate the reflection color of every single microdroplets independently, and Figure 2.10g–h illustrates the "flower-opening" patterns with light-directed iridescent colors in eight monodisperse microdroplets. Such monodisperse microdroplets and their periodic arrays would not only provide a rich and fascinating platform for fundamental theoretical studies of micro/nanophotonics with geometric confinements but also hold great potential for applications in the photonic and optoelectronic devices where CPL with tailored wavelength is preferred.[98,99]

The microfluidic techniques have also been demonstrated to fabricate the CLC microshells with spherical configurations that can act as 3D omnidirectional lasers due to the radial molecular arrangement within the cholesteric microshells. Uchida et al. firstly reported the fabrication of cholesteric microshells with water–oil–water double phase by using the microfluidic method,[100] in which the oil phase is composed of dye-doped CLCs. By combining different laser dyes with CLC hosts, three types of laser modes, that is, distributed feedback (DFB), distributed Bragg reflection (DBR) and whispering-gallery-mode (WGM) resonators, are successfully obtained

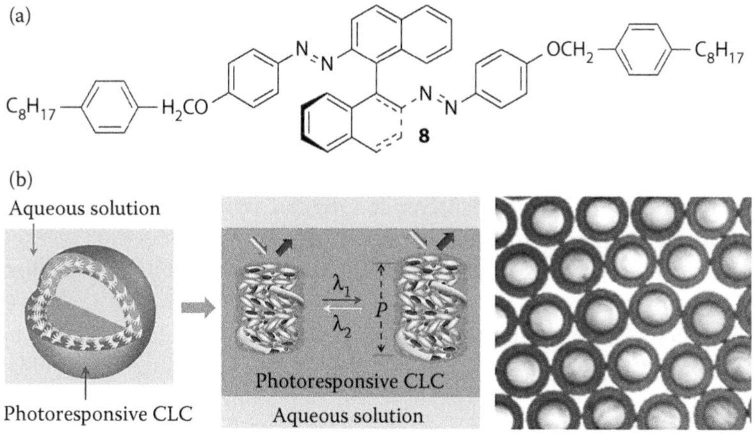

FIGURE 2.11 (a) Molecular structure of light-driven chiral azobenezene **8**. (b) Schematic illustrations of the cholesteric microshell and the mechanism of phototunable lasing (left) and confocal image of the microshells (right). (Chen, L. et al.: *Adv. Opt. Mater.* 845–848. 2014. Copyright Wiley-VCH Verlag GmbH & Co. KGaA. Reproduced with permission.)

on the basis of the CLC microshells. Subsequently, we demonstrated the fabrication and lasing properties of photochromic monodisperse cholesteric liquid crystalline microshells through the capillary-based microfluidic techniques (Figure 2.11),[101] where the photoresponsive CLCs were developed by loading a photochromic chiral molecular switch **8** responsive to visible light into a nematic LC host. When the cholesteric microshells loaded with DCM laser dye were pumped under an external pulsed laser, the laser emissions in all optical directions were achieved, where the emission wavelength was found to be able to self-tune because of the photoisomerization of azobenzene chiral switch.

2.5 PHOTOCHROMIC LIQUID CRYSTALS FOR POTENTIAL INFRARED SENSORS

As is well known, most infrared radiations are not visible to human eye and may harm a person without warning, making it difficult for us to avoid the risk. Although occasional exposure to these low-energy radiations is not harmful, everyday contact can result in long-term health problems. People who are involved in the mass production of lasers and heat lamps could have a higher risk of experiencing adverse effects due to exposure. Prolonged exposure to high levels of infrared radiation could lead to burns and permanent eye problems, including cornea and retina damage, cataracts, and other injuries to eye.[102] Toward the fabrication of infrared photochromic sensors, extensive efforts were attempted to design chiral molecular switches that can be triggered with longer wavelengths including visible light.[103–105] However, developing such infrared photochromic molecules continues to be a challenge but is attracting much attention due to their fundamental significance and potential technological application.

Upconversion processes are characterized by the successive absorption of multiple photons and the subsequent emission of shorter wavelength radiation.[106,107] Thus, the systems exploiting this phenomenon could be used to first absorb several quanta of infrared light and subsequently transfer their upconverted excitation energy to the photochromic switch either via UV emission and reabsorption by the switch or Förster-type resonance energy transfer. Extensive research have indicated that nanoparticles doped with lanthanide ions can be used to drive the isomerization of photochromic compounds in one direction, whereas complex core–shell–shell nanostructures are necessary to photoisomerize in both directions.[108–111] Recently, we introduced the reversible NIR-directed reflection in a self-organized helical superstructure loaded with upconversion nanoparticles (UCNPs) possessing core-shell structure as shown in Figure 2.12.[112] The photochromic chiral azobenzene switch **9** and UCNPs, doped in a LC medium, were able to self-organize into an optically tunable helical superstructure. The resulting nanoparticle impregnated helical superstructure was found to exhibit unprecedented reversible NIR light-activated tunable behavior only by modulating the excitation power density of a continuous-wave NIR laser. Interestingly, upon irradiation by the NIR laser at the high power density, the reflection wavelength of the photonic superstructure redshifted, whereas its reverse process occurred upon irradiation by the same laser but with the lower power density.

In the experiment, the self-organized helical superstructure was fabricated by doping a photchromic chiral azobenzene switch **9** and UCNPs into a LC host

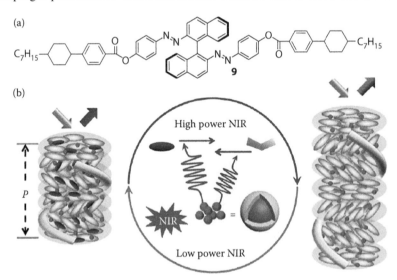

Reversible NIR-light directed reflection

FIGURE 2.12 (a) Molecular structure of light-driven chiral azobenezene **9**. (b) Upconversion-nanoparticle-doped reflection tuning of photochromic chiral LC. (Reprinted with permission from Wang, L. et al. Reversible near-infrared light directed reflection in a self-organized helical superstructure loaded with upconversion nanoparticles. *J. Am. Chem. Soc.* 136, 4480–4483. Copyright 2014 American Chemical Society.)

composed of commercially available components. It was found that 3 wt% chiral switch and 1.5 wt% UCNPs doped in commercially available 80.5 wt% nematic LC E7 and 15 wt% S811 exhibited reversibly tunable NIR light-directed reflections through red, green, and blue wavelength only by varying the power density of the excitation NIR light (Figure 2.13a and b). As shown in Figure 2.13c, the center reflection wavelength was around 435 nm at the initial state; upon irradiation with 980-nm NIR laser at high power density, the reflection wavelength was tuned to 540 nm in 60 s and further reached a photostationary state in 120 s with a center reflection wavelength at 625 nm. The reverse process across red, green, and blue reflection colors occurred within ~4 min upon irradiation with 980-nm NIR laser at low power density (Figure 2.13d). The reversible tuning of reflection across red, green, and blue reflection colors was repeated many times without noticeable degradation.

Recently, plasmonic gold nanorods (GNRs) have attracted increasing attentions especially in biomedical applications due to their appealing "photothermal effect," that is, converting NIR light into heat through a nonradiative relaxation

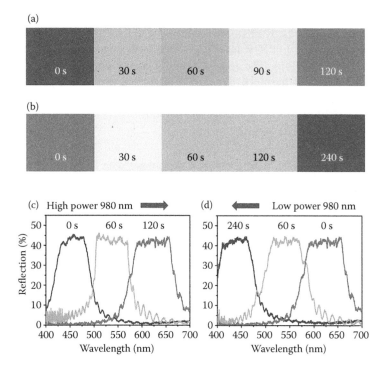

FIGURE 2.13 Reflection optical images and corresponding reflection spectra of the photochromic chiral LC with 3 wt% chiral molecular switch **9** and 1.5 wt% UCNPs in a 10-μm thick planar cell at room temperature: (a) and (c) upon irradiation with 980-nm NIR laser at high power density and followed by (b) and (d) irradiation with 980-nm NIR laser at low power density. Note: For the colorful images, please check the original figures in the reference. (Reprinted with permission from Wang, L. et al. Reversible near-infrared light directed reflection in a self-organized helical superstructure loaded with upconversion nanoparticles. *J. Am. Chem. Soc.* 136, 4480–4483. Copyright 2014 American Chemical Society.)

process of longitudinal surface plasmon resonance (LSPR).[113] It is also noteworthy in this context that the characteristic wavelength of the LSPR signal corresponding to NIR absorption can be expediently tuned on demand by suitably altering the aspect ratio of GNRs, and therefore GNRs have been proposed for use in a few triggering systems where UV or visible light is not favorable.[114–117] We demonstrated NIR-light-responsive self-organized 3D photonic superstructure by incorporating new hydrophobic mesogen-functionalized GNRs (M-GNRs) into a BP LC medium composed of commercially available components.[118] In the experiment, the liquid crystalline composites doped with 0.07 wt% concentration of M-GNRs were prepared by dispersing the M-GNRs into BP LCs. On cooling from isotropic phase, BP II firstly appeared at 46.4°C and then transformed into BP I at 44.1°C. As the temperature was further decreased to 37.2°C, the phase completely changed into the cholesteric phase (Figure 2.14a). Additionally, the BPs formed the characteristic platelet textures, and the reflection color gradually changed to a longer wavelength with decreasing temperature. It was also found that the introduction of M-GNRs was found to be beneficial in stabilizing the cubic nanostructure. Importantly, the resultant 3D photonic nanostructures could be switched between body-centered cubic and simple cubic symmetry under irradiation using an 808-nm NIR laser due to the significant photothermal effect of M-GNRs (Figure 2.14b). The reverse process occurs upon removal of the NIR laser irradiation. Furthermore, dynamic NIR-light-directed red, green, and blue (RGB) reflections were for the first time demonstrated by tuning the lattice constant of the light-driven 3D soft photonic crystals.

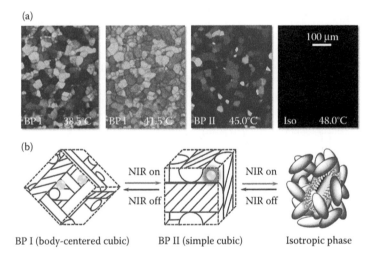

FIGURE 2.14 (See color insert.) NIR-light-directed self-organized BP 3D photonic superstructures loaded with gold nanorods. (a) Typical textures of BP I and BP II nanostructures at different temperatures and (b) schematic illustration of the structural transformations under the NIR-light irradiation. (Wang, L. et al. 2015. NIR light-directing self-organized 3D photonic superstructures loaded with anisotropic plasmonic hybrid nanorods. *Chem. Comm.* 51: 15039–15042. Reproduced by permission of The Royal Society of Chemistry.)

2.6 SUMMARY AND PERSPECTIVES

In this chapter, we furnished a glimpse of the research advances in design and development of photochromic chiral LCs for potential light sensing applications, that is, light-driven dynamic circularly polarized reflections of photochromic cholesteric and BP LCs. The combination of photochromic behavior of molecular switches with dynamic circularly polarized visible reflection of multitudinous chiral LCs helps the human eye to directly "see" the light radiations and even "feel" their intensities. This kind of smart sensors possessing unique attributes such as light-weight, low cost, and easy-to-use hold great potential for being incorporated in the tags or packages of perishable products and any other portable daily items, which would help to increase the public awareness of these hazardous radiations such as UV, infrared exposures, and beyond. However, it should be pointed out that the related research in this field is still in infancy stage which is far from being usable in daily life situations. Often the selectivity, sensitivity, and signal changes do not meet the requirements set for the specific applications. To increase the sensitivity of photochromic chiral LC sensors, more attentions should be paid to design and synthesis of novel photochromic molecules with unprecedented fast switching speeds and remarkable stabilities.[119] Toward the photochromic sensors, the linear relationship between the reflection colors of chiral LCs and the parameters of light irradiations including wavelength, intensity, polarization, and exposure time requires detail investigation in the future.

As aforementioned, most of photochromic chiral LCs reported so far are based on low-molecular-weight systems in which a promising photochromic chiral molecule is used. However, these samples have a liquid-like nature which often hampers their widespread application. To develop the paintable, flexible, and even wearable smart sensors, it would be a great choice to encapsulate the photochromic chiral LCs with a polymer binder to create polymer-dispersed LCs in which the good response of low-weight systems can be combined with the mechanical robustness of polymers. Furthermore, it would be of great interest to develop intelligent fibers, fabrics, textiles, or even clothes by directly using the photochromic chiral LCs.[120–122] This kind of textile-based photochromic sensors should be easily customizable by sewing, thermal bonding, or gluing, and there are also additional advantages of easy maintenance through washing and chemical cleaning, and low specific weight with good strength and elasticity. Looking forward, although extensive advancements in developing the photochromic sensors based on LCs have been achieved, various challenges remain to construct those displaying high sensitivities and stabilities for the general public use. Future investigations in this promising topic with great potential would not only broaden our knowledge of smart sensors but also promote their diverse applications in daily life situations.

ACKNOWLEDGMENTS

The preparation of this chapter benefited from the support to Quan Li by U.S. Air Force Office of Scientific Research (AFOSR), U.S. National Aeronautics and Space Administration (NASA) and U.S. National Science Foundation (NSF), U.S. Army Research Office (ARO), U.S. Department of Defense Multidisciplinary University

Research Initiative (DoD MURI), U.S. Department of Energy (DOE), and Ohio Third Frontier. We thank all the Li's lab current and former members as well as his collaborators, whose names are found in the references, for their significant contributions in this project.

REFERENCES

1. Lang, K. R. 1995. *Sun, Earth and Sky*. Springer-Verlag, New York.
2. Melnikova, I. N. and A. V. Vasilyev. 2005. *Short-Wave Solar Radiation in the Earth's Atmosphere: Calculation, Observation, Interpretation*. Springer-Verlag, Berlin, Heidelberg.
3. Diffey, B. L. 1991. Solar ultraviolet radiation effects on biological systems. *Phys. Med. Biol.* 36: 299–328.
4. Berger, D. S. 1976. The sunburning ultraviolet meter: Design and performance. *Photochem. Photobiol.* 24: 587–593.
5. Konstantatos, G. and E. H. Sargent. 2010. Nanostructured materials for photon detection. *Nature Nanotech.* 5: 391–400.
6. Bogue, R. 2009. Nanosensors: A review of recent research. *Sensor Rev.* 29: 310–315.
7. Funakoshi, N., F. Ebisawa, M. Hoshino, T. Yoshida, K. Sukegawa, A. Morinaka, N. Sashida, S. Toeda, and M. Urabe. 1995. Photochromic compound. U.S. Pat. No. 5387798.
8. Goudjil, K. 1996. Photochromic ultraviolet detector. U.S. Pat. No. 5591090.
9. Qin, M., Y. Huang, F. Li, and Y. Song. 2015. Photochromic sensors: A versatile approach for recognition and discrimination. *J. Mater. Chem. C* 3: 9265–9275.
10. Goudjil, K. and R. Sandoval. 1998. Photochromic ultraviolet light sensor and applications. *Sensor Rev.* 18: 176–177.
11. Chen, R. H. 2011. *Liquid Crystal Displays: Fundamental Physics and Technology*. John Wiley & Sons, Hoboken, New Jersey.
12. Li, Q. (Ed.). 2012. *Liquid Crystals Beyond Displays: Chemistry, Physics, and Applications*. John Wiley & Sons, Hoboken, New Jersey.
13. Bisoyi, H. and Q. Li, 2014. *Liquid Crystals in Kirk-Othmer Encyclopedia of Chemical Technology*. John Wiley & Sons, 1–52. http://onlinelibrary.wiley.com/doi/10.1002/0471 238961.1209172103151212.a01.pub3/abstract
14. Montbach, E., N. Venkataraman, J. W. Doane, A. Khan, G. Magyar, I. Shiyanovskaya, T. Schneider, L. Green, and Q. Li. 2008. Novel optically addressable photochiral displays. *SID Digest Tech. Papers* 39: 919–922.
15. Oswald P. and P. Pieranski. 2005. *Nematic and Cholesteric Liquid Crystals: Concepts and Physical Properties Illustrated by Experiments*. Taylor & Francis, CRC Press, Boca Raton, Florida.
16. Green, L., Y. Li, T. White, A. Urbas, T. Bunning, and Q. Li. 2009. Light-driven chiral molecular switches with tetrahedral and axial chirality. *Org. Biomol. Chem.* 7: 3930–3933.
17. Mathews, M., R. Zola, D. Yang, and Q. Li. 2011. Thermally, photochemically and electrically switchable reflection colors from self-organized chiral bent-core liquid crystals. *J. Mater. Chem.* 21: 2098–2103.
18. Wang, Y., Z. Zheng, H. K. Bisoyi, K. G. Gutierrez-Cuevas, L. Wang, R. S. Zola, and Q. Li. 2016. Thermally reversible full color selective reflection in a self-organized helical superstructure enabled by a bent-core oligomesogen exhibiting twist-bend nematic phase. *Mater. Horiz.* 3: 442–446.
19. Xie, H., L. Wang, H. Wang, C. Zhou, M. Wang, B. Wang, Z. Chen, L. Zhang, X. Zhang, and H. Yang. 2015. Electrically tunable properties of wideband-absorptive and reflection-selective films based on multi-dichroic dye-doped cholesteric liquid crystals. *Liq. Cryst.* 42: 1698–1705.

20. Salili, S. M., J. Xiang, H. Wang, Q. Li, D. A. Paterson, J. M. D. Storey, C. T. Imrie, O. D. Lavrentovich, S. N. Sprunt, J. T. Gleeson, and A. Jákli. 2016. Magnetically tunable selective reflection of light by heliconical cholesterics. *Phys. Rev. E* 94: 042705.
21. Gutierrez-Cuevas, K. G., L. Wang, Z. Zheng, H. K. Bisoyi, G. Li, L. S. Tan, R. Vaia, and Q. Li. 2016. Frequency-driven self-organized helical superstructures loaded with mesogen-grafted silica nanoparticles. *Angew. Chem. Int. Ed. Engl.* 55: 13090–13094.
22. Li, Q. (Ed.). 2013. *Intelligent Stimuli-Responsive Materials: From Well-Defined Nanostructures to Applications.* John Wiley & Sons, Hoboken, New Jersey.
23. Wang, L. and Q. Li. 2015. Stimuli responsive self-organized liquid crystalline nanostructures: From 1D to 3D photonic crystals. In *Organic & Hybrid Photonic Crystals.* D. Comoretto (Ed.), Springer, New York, pp. 393–430, Chapter 18.
24. Carlton, R. J., J. T. Hunter, D. S. Miller, R. Abbasi, P. C. Mushenheim, L. N. Tan, and N. L. Abbott. 2013. Chemical and biological sensing using liquid crystals. *Liq. Cryst. Rev.* 1: 29–51.
25. Chanishvili, A., G. Chilaya, G. Petriashvili, R. Barberi, R. Bartolino, and M. P. De Santo. 2005. Cholesteric liquid crystal mixtures sensitive to different ranges of solar UV irradiation. *Mol. Cryst. Liq. Cryst.* 434: 25–353.
26. Petriashvili, G., A. Chanishvili, G. Chilaya, M. A. Matranga, M. P. De Santo, and R. Barberi. 2009. Novel UV sensor based on a liquid crystalline mixture containing a photoluminescent dye. *Mol. Cryst. Liq. Cryst.* 500: 82–90.
27. Mulder, D. J., A. P. H. J. Schenning, and C. W. M. Bastiaansen. 2014. Chiral-nematic liquid crystals as one dimensional photonic materials in optical sensors. *J. Mater. Chem. C* 2: 6695–6705.
28. Han, Y., K. Pacheco, C. W. M. Bastiaansen, D. J. Broer, and R. P. Sijbesma. 2010. Optical monitoring of gases with cholesteric liquid crystals. *J. Am. Chem. Soc.* 132: 2961–2967.
29. Herzer, N., H. Guneysu, D. J. D. Davies, D. Yildirim, A. R. Vaccaro, D. J. Broer, C. W. M. Bastiaansen, and A. P. H. J. Schenning. 2012. Printable optical sensors based on H-bonded supramolecular cholesteric liquid crystal networks. *J. Am. Chem. Soc.* 134: 7608–7611.
30. Su, X., S. Voskian, R. P. Hughes, and I. Aprahamian. 2013. Manipulating liquid-crystal properties using a pH activated hydrazone switch. *Angew. Chem. Int. Ed. Engl.* 125: 10934–10939.
31. Zhang, J., Q. Zou, and H. Tian. 2013. Photochromic materials: More than meets the eye. *Adv. Mater.* 25: 378–399.
32. Tamaoki, N. and T. Kamei. 2010. Reversible photo-regulation of the properties of liquid crystals doped with photochromic compounds. *J. Photochem. Photobiol. C: Photochem. Rev.* 11: 47–61.
33. Wang, Y. and Q. Li. 2012. Light-driven chiral molecular switches or motors in liquid crystals. *Adv. Mater.* 24: 1926–1945.
34. Irie, M. 2000. Diarylethenes for memories and switches. *Chem. Rev.* 100: 1685–1716.
35. Irie, M., T. Fukaminato, K. Matsuda, and S. Kobatake. 2014. Photochromism of diarylethene molecules and crystals: Memories, switches, and actuators. *Chem. Rev.* 114: 12174–12277.
36. Minkin, V. I. 2004. Photo-, thermo-, solvato-, and electrochromic spirohetero-cyclic compounds. *Chem. Rev.* 104: 2751–2776.
37. Berkovic, G., V. Krongauz, and V. Weiss. 2000. Spiropyrans and spirooxazines for memories and switches. *Chem. Rev.* 100: 1741–1754.
38. Kawata, S. and Y. Kawata. 2000. Three-dimensional optical data storage using photochromic materials. *Chem. Rev.* 100: 1777–1788.
39. Minkin, V. I. 2011. Photoswitchable molecular systems based on spiropyrans and spirooxazines. *Mol. Switch.* Second Edition: 37–80.

40. Guram, C. 2001. Cholesteric liquid crystals: Optics, electro-optics, and photo-optics. In *Chirality in Liquid Crystals*. H.-S. Kitzerow, C. Bahr (Eds.), Springer, New York, pp. 159–185.

41. Li, Y. and Q. Li. 2014. Photoresponsive chiral liquid crystal materials: From 1D helical superstructures to 3D periodic cubic lattices and beyond. In *Nanoscience with Liquid Crystals: From Self-Organized Nanostructures to Applications*. Q. Li (Ed.), Springer, Heidelberg, pp. 135–177, Chapter 5.

42. Lin, T.-H., C.-W. Chen, and Q. Li. 2015. Self-organized 3D photonic superstructure: Blue phase liquid crystal. In *Anisotropic Nanomaterials: Preparation, Properties, and Applications*. Q. Li (Ed.), Springer, Heidelberg, pp. 337–378, Chapter 9.

43. Wang, L., W. He, X. Xiao, F. Meng, Y. Zhang, P. Yang, L. Wang, J. Xiao, H. Yang, and Y. Lu. 2012. Hysteresis-free blue phase liquid-crystal-stabilized by ZnS nanoparticles. *Small* 8: 2189–2193.

44. Coles, H. J. and M. N. Pivnenko. 2005. Liquid crystal "blue phases" with a wide temperature range. *Nature* 436: 997–1000.

45. Wang, L., W. He, X. Xiao, Q. Yang, B. Li, P. Yang, and H. Yang. 2012. Wide blue phase range and electro-optical performances of liquid crystalline composites doped with thiophene-based mesogens. *J. Mater. Chem.* 22: 2383–2386.

46. Kikuchi, H., M. Yokota, Y. Hisakado, H. Yang, and T. Kajiyama. 2002. Polymer-stabilized liquid crystal blue phases. *Nat. Mater.* 1: 64–68.

47. Li, B., W. He, L. Wang, X. Xiao, and H. Yang. 2013. Effect of lateral fluoro substituents of rodlike tolane cyano mesogens on blue phase temperature ranges. *Soft Matter* 9: 1172–1177.

48. Wang, L., W. He, M. Wang, M. Wei, J. Sun, X. Chen, and H. Yang. 2013. Effects of symmetrically 2, 5-disubstituted 1, 3, 4-oxadiazoles on the temperature range of liquid crystalline blue phases: A systematic study. *Liq. Cryst.* 40: 354–367.

49. Wang, L., L. Yu, X. Xiao, Z. Wang, P. Yang, W. He, and H. Yang. 2012. Effects of 1, 3, 4-oxadiazoles with different rigid cores on the thermal and electro-optical performances of liquid crystalline blue phases. *Liq. Cryst.* 39: 629–638.

50. Wang, L., W. He, Q. Wang, M. Yu, X. Xiao, Y. Zhang, M. Ellahi, D. Zhao, H. Yang, and L. Guo. 2013. Polymer-stabilized nanoparticle-enriched blue phase liquid crystals. *J. Mater. Chem. C* 1: 6526–6531.

51. Wang, L., W. He, X. Xiao, M. Wang, M. Wang, P. Yang, Z. Zhou, H. Yang, H. Yu, and Y. Lu. 2012. Low voltage and hysteresis-free blue phase liquid crystal dispersed by ferroelectric nanoparticles. *J. Mater. Chem.* 22: 19629–19633.

52. Cao, W., A. Muñoz, P. Palffy-Muhoray, and B. Taheri. 2002. Lasing in a three-dimensional photonic crystal of the liquid crystal blue phase II. *Nat. Mater.* 1: 111–113.

53. Coles, H. and S. Morris. 2010. Liquid-crystal lasers. *Nat. Photonics* 4: 676–685.

54. Zheng, Z., Y. Li, H. K. Bisoyi, L. Wang, T. J. Bunning, and Q. Li. 2016. Three-dimensional control of the helical axis of a chiral nematic liquid crystal by light. *Nature* 531: 352–356.

55. Bisoyi, H. K. and Q. Li. 2016. Light-driven liquid crystalline materials: From photo-induced phase transitions and property modulations to applications. *Chem. Rev.* 116: 15089–15166.

56. Chen, X., L. Wang, Y. Chen, C. Li, G. Hou, X. Liu, X. Zhang, W. He, and H. Yang. 2014. Broadband reflection of polymer-stabilized chiral nematic liquid crystals induced by a chiral azobenzene compound. *Chem. Comm.* 50: 691–694.

57. Zheng, Z. and Q. Li. 2016. Self-organized chiral liquid crystalline nanostructures for energy-saving devices. In *Nanomaterials for Sustainable Energy*. Q. Li (Ed.), Springer, Heidelberg, pp. 513–554, Chapter 14.

58. Bisoyi, H. K. and Q. Li. 2016. Light-directed dynamic chirality inversion in functional self-organized helical superstructures. *Angew. Chem. Int. Ed. Engl.* 55: 2994–3010.

59. Eelkema, R. and B. L. Feringa. 2006. Amplification of chirality in liquid crystals. *Org. Biomol. Chem.* 4: 3729–3745.

60. Pieraccini, S., S. Masiero, A. Ferrarini, and G. P. Spada. 2011. Chirality transfer across length-scales in nematic liquid crystals: Fundamentals and applications. *Chem. Soc. Rev.* 40: 258–271.

61. Li, Y., M. Wang, A. Urbas, and Q. Li. 2013. Photoswitchable but thermally stable axially chiral dithienylperfluorocyclopentene dopant with high helical twisting power. *J. Mater. Chem. C* 1: 3889–4044.

62. Mathews, M., R. Zola, Hurley, S., D. Yang, T. J. White, T. J. Bunning, and Q. Li. 2010. Light-driven reversible handedness inversion in self-organized helical superstructures. *J. Am. Chem. Soc.* 132: 18361–18366.

63. Li, Y., M. Wang, T. J. White, T. J. Bunning, and Q. Li. 2013. Azoarenes bearing opposite chiral configurations: Light-driven dynamic reversible handedness inversion in self-organized helical superstructure. *Angew. Chem. Int. Ed. Engl.* 52: 8925–8929.

64. Sackmann, E. 1971. Photochemically induced reversible color changes in cholesteric liquid crystals. *J. Am. Chem. Soc.* 93: 7088–7090.

65. Mathews, M. and N. Tamaoki. 2008. Planar chiral azobenzenophanes as chiroptic switches for photon mode reversible reflection color control in induced chiral nematic liquid crystals. *J. Am. Chem. Soc.* 130: 11409–11416.

66. Tamaoki, N. 2001. Cholesteric liquid crystals for color information technology. *Adv. Mater.* 13: 1135–1147.

67. Brehmer, M., J. Lub, and P. van de Witte. 1998. Light-induced color change of cholesteric copolymers. *Adv. Mater.* 10: 1438–1441.

68. Yoshioka, T., T. Ogata, T. Nonaka, M. Moritsugu, S.-N. Kim, and S. Kurihara. 2005. Reversible-photon-mode full-color display by means of photochemical modulation of a helically cholesteric structure. *Adv. Mater.* 17: 1226–1229.

69. Pieraccini, S., S. Masiero, G. P. Spada, and G. Gottarelli. 2003. A new axially chiral photochemical switch. *Chem. Commun.* 5: 598–599.

70. Li, Q., L. Green, N. Venkataraman, I. Shiyanovskaya, A. Khan, A. Urbas, and J. W. Doane. 2007. Reversible photoswitchable axially chiral dopants with high helical twisting power. *J. Am. Chem. Soc.* 129: 12908–12909.

71. Ma, J., Y. Li, T. White, A. Urbas, and Q. Li. 2010. Light-driven nanoscale chiral molecular switch: Reversible dynamic full range color phototuning. *Chem. Commun.* 46: 3463–3465.

72. Li, Q., Y. Li, J. Ma, D. K. Yang, T. J. White, and T. J. Bunning. 2011. Directing dynamic control of red, green, and blue reflection enabled by a light-driven self-organized helical superstructure. *Adv. Mater.* 23: 5069–5073.

73. Denekamp, C. and B. L. Feringa. 1998. Optically active diarylethenes for multimode photoswitching between liquid-crystalline phases. *Adv. Mater.* 10: 1080–1082.

74. Yamaguchi, T., T. Inagawa, H. Nakazumi, S. Irie, and M. Irie. 2000. Photoswitching of helical twisting power of a chiral diarylethene dopant: Pitch change in a chiral nematic liquid crystal. *Chem. Mater.* 12: 869–871.

75. Li, Y., A. Urbas, and Q. Li. 2011. Synthesis and characterization of light-driven dithienylcyclopentene switches with axial chirality. *J. Org. Chem.* 76: 7148–7156.

76. Li, Y., A. Urbas, and Q. Li. 2012. Reversible light-directed red, green, and blue reflection with thermal stability enabled by a self-organized helical superstructure. *J. Am. Chem. Soc.* 134: 9573–9576.

77. Li, Y., C. Xue, M. Wang, A. Urbas, and Q. Li. 2013. Photodynamic chiral molecular switches with thermal stability: From reflection wavelength tuning to handedness inversion of self-organized helical superstructures. *Angew. Chem. Int. Ed. Engl.* 52: 13703–13707.

78. Bisoyi, H. K. and Q. Li. 2014. Light-directing chiral liquid crystal nanostructures: From 1D to 3D. *Acc. Chem. Res.* 47: 3184–3195.

79. Hattori, H. and T. Uryu. 2001. Photochromic chiral liquid crystalline systems containing spiro-oxazine with a chiral substituent II. Photoinduced behaviour. *Liq. Cryst.* 28: 1099–1104.

80. Jin, L., Y. Li, J. Ma, and Q. Li. 2010. Synthesis of novel thermally reversible photochromic axially chiral spirooxazines. *Org. Lett.* 12: 3552–3555.

81. van Delden, R. A., M. B. van Gelder, N. P. M. Huck, and B. L. Feringa. 2003. Controlling the color of cholesteric liquid-crystalline films by photoirradiation of a chiroptical molecular switch used as dopant. *Adv. Funct. Mater.* 13: 319–324.

82. van Delden, R. A., N. Koumura, N. Harada, and B. L. Feringa. 2002. Supramolecular chemistry and self-assembly special feature: Unidirectional rotary motion in a liquid crystalline environment: Color tuning by a molecular motor. *Proc. Natl. Acad. Sci. USA* 99: 4945–4949.

83. Eelkema, R. and B. L. Feringa. 2006. Reversible full-range color control of a cholesteric liquid crystalline film by using a molecular motor. *Chem. Asian J.* 1: 367–369.

84. Aßhoff, S. J., S. Iamsaard, A. Bosco, J. J. L. M. Cornelissen, B. L. Feringa, and N. Katsonis. 2013. Time-programmed helix inversion in phototunable liquid crystals. *Chem. Commun.* 49: 4256–4258.

85. Wang, L. and Q. Li. 2015. Stimuli-directing self-organized 3D liquid crystalline nanostructures: From materials design to photonic applications. *Adv. Funct. Mater.* 26: 10–28.

86. Nishijima, Y., K. Ueno, S. Juodkazis, V. Mizeikis, H. Misawa, T. Tanimura, and K. Maeda. 2007. Inverse silica opal photonic crystals for optical sensing applications. *Optics Express* 15: 12979–12988.

87. Baryshev, A., R. Fujikawa, A. Khanikaev, A. Granovsky, K. H. Shin, P. B. Lim, and M. Inoue. 2006. Mesoporous photonic crystals for sensor applications. *Proc. of SPIE* 6369: 63690B.

88. Chen, X., L. Wang, C. Li, J. Xiao, H. Ding, X. Liu, X. Zhang, W. He, and H. Yang. 2013. Light-controllable reflection wavelength of blue phase liquid crystals doped with azobenzene-dimers. *Chem. Commun.* 49: 10097–10099.

89. Chanishvili, A., G. Chilaya, G. Petriashvili, and P. J. Collings. 2005. Trans-cis isomerization and the blue phases. *Phys. Rev. E* 71: 051705.

90. Liu, H. Y., C. T. Wang, C. Y. Hsu, T. H. Lin, and J. H. Liu. 2010. Optically tuneable blue phase photonic band gaps. *Appl. Phys. Lett.* 96: 121103.

91. Lin, T. H., Y. Li, C. T. Wang, H. C. Jau, C. W. Chen, C. C. Li, H. K. Bisoyi, T. J. Bunning, and Q. Li. 2013. Red, green and blue reflections enabled in an optically tunable self-organized 3D cubic nanostructured thin film. *Adv. Mater.* 25: 5050–5054.

92. Yang, D. K. and P. P. Crooker. 1987. Chiral-racemic phase diagrams of blue-phase liquid crystals. *Phys. Rev. A* 35: 4419–4423.

93. Cipparrone, G., A. Mazzulla, A. Pane, R. J. Hernandez, and R. Bartolino. 2011. Chiral self-assembled solid microspheres: A novel multifunctional microphotonic device. *Adv. Mater.* 23: 5773–5778.

94. Utada, A. S., E. Lorenceau, D. R. Link, P. D. Kaplan, H. A. Stone, and D. A. Weitz. 2005. Monodisperse double emulsions generated from a microcapillary device. *Science* 308: 537–541.

95. Lee, H. G., S. Munir, and S. Y. Park. 2016. Cholesteric liquid crystal droplets for biosensors. *ACS Appl. Mater. Interfaces* 8: 26407–26417.

96. Geng, Y., J. H. Noh, I. Drevensek-Olenik, R. Rupp, G. Lenzini, and J. P. F. Lagerwall. 2016. High-fidelity spherical cholesteric liquid crystal Bragg reflectors generating unclonable patterns for secure authentication. *Sci. Rep.* 6: 26840.

97. Fan, J., Y. Li, H. K. Bisoyi, R. S. Zola, D. K. Yang, T. J. Bunning, D. A. Weitz, and Q. Li. 2015. Light-directing omnidirectional circularly polarized reflection from liquid-crystal droplets. *Angew. Chem. Int. Ed. Engl.* 127: 2188–2192.

98. Noh, J. H., H. L. Liang, I. Drevensek-Olenik, and J. P. F. Lagerwall. 2014. Tuneable multicoloured patterns from photonic cross-communication between cholesteric liquid crystal droplets. *J. Mater. Chem. C* 2: 806–810.

99. Lee, S. S., B. Kim, S. K. Kim, J. C. Won, Y. H. Kim, and S. H. Kim. 2015. Robust microfluidic encapsulation of cholesteric liquid crystals toward photonic ink capsules. *Adv. Mater.* 27: 627–633.

100. Uchida, Y., Y. Takanishi, and J. Yamamoto. 2013. Controlled fabrication and photonic structure of cholesteric liquid crystalline shells. *Adv. Mater.* 25: 3234–3237.

101. Chen, L., Y. Li, J. Fan, H. K. Bisoyi, D. A. Weitz, and Q. Li. 2014. Photoresponsive monodisperse cholesteric liquid crystalline microshells for tunable omnidirectional lasing enabled by a visible light-driven chiral molecular switch. *Adv. Opt. Mater.* 2: 845–848.

102. Cho, S., M. H. Shin, Y. K. Kim, J. E. Seo, Y. M. Lee, C. H. Park, and J. H. Chung. 2009. Effects of infrared radiation and heat on human skin aging in vivo. J. Investig. *Dermatol. Symp. Proc.* 14: 15–19.

103. Wang, Y., A. Urbas, and Q. Li. 2012. Reversible visible-light tuning of self-organized helical superstructures enabled by unprecedented light-driven axially chiral molecular switches. *J. Am. Chem. Soc.* 134: 3342–3345.

104. Yang, Y., R. P. Hughes, and I. Aprahamian. 2014. Near-infrared light activated azo-BF$_2$ Switches. *J. Am. Chem. Soc.* 36: 13190–13193.

105. Bléger, D. and S. Hecht. 2015. Visible light activated molecular switches. *Angew. Chem. Int. Ed. Engl.* 54: 2–14.

106. Wang, F. and X. Liu. 2009. Recent advances in the chemistry of lanthanide-doped upconversion nanocrystals. *Chem. Soc. Rev.* 38: 976–989.

107. Dong, H., L.-D. Sun, and C.-H. Yan. 2015. Energy transfer in lanthanide upconversion studies for extended optical applications. *Chem. Soc. Rev.* 44: 1608–1634.

108. Carling, C. J., J. C. Boyer, and N. R. Branda. 2009. Remote-control photoswitching using NIR light. *J. Am. Chem. Soc.* 131: 10838–10839.

109. Boyer, J. C., C. J. Carling, B. D. Gates, and N. R. Branda. 2010. Two-way photoswitching using one type of near-infrared light, upconverting nanoparticles, and changing only the light intensity. *J. Am. Chem. Soc.* 132: 15766–15772.

110. Wu, W., L. Yao, T. Yang, R. Yin, F. Li, and Y. Yu. 2011. NIR-light-induced deformation of cross-linked liquid-crystal polymers using upconversion nanophosphors. *J. Am. Chem. Soc.* 133: 15810–15813.

111. Wang, L., H. Dong, Y. Li, R. Liu, Y. F. Wang, H. K. Bisoyi, L. D. Sun, C. H. Yan, and Q. Li. 2015. Luminescence-driven reversible handedness inversion of self-organized helical superstructures enabled by a novel near-infrared light nanotransducer. *Adv. Mater.* 27: 2065–2069.

112. Wang, L., H. Dong, Y. Li, C. Xue, L.-D. Sun, C.-H. Yan, and Q. Li. 2014. Reversible near-infrared light directed reflection in a self-organized helical superstructure loaded with upconversion nanoparticles. *J. Am. Chem. Soc.* 136: 4480–4483.

113. Chen, H., L. Shao, Q. Li, and J. Wang. 2013. Gold nanorods and their plasmonic properties. *Chem. Soc. Rev.* 42: 2679–2724.

114. Gorelikov, I., L. M. Field, and E. Kumacheva. 2004. Hybrid microgels photoresponsive in the near-infrared spectral range. *J. Am. Chem. Soc.* 126: 15938–15939.

115. Xue, C., J. Xiang, H. Nemati, H. K. Bisoyi, K. Gutierrez-Cuevas, L. Wang, M. Gao et al. 2015. Light-driven reversible alignment switching of liquid crystals enabled by azo thiol grafted gold nanoparticles. *Chemphyschem.* 16: 1852–1856.

116. Wang, L., K. G. Gutierrez-Cuevas, A. Urbas, and Q. Li. 2015. Near-infrared light-directed handedness inversion in plasmonic nanorod embedded helical superstructure. *Adv. Opt. Mater.* 4: 247–251.

117. Gutierrez-Cuevas, K. G., L. Wang, C. Xue, G. Singh, S. Kumar, A. Urbas, and Q. Li. 2015. Near infrared light-driven liquid crystal phase transition enabled by hydrophobic mesogen grafted plasmonic gold nanorods. *Chem. Comm.* 51: 9845–9848.

118. Wang, L., K. G. Gutierrez-Cuevas, H. K. Bisoyi, J. Xiang, G. Singh, R. S. Zola, S. Kumar, O. D. Lavrentovich, A. Urbas, and Q. Li. 2015. NIR light-directing self-organized 3D photonic superstructures loaded with anisotropic plasmonic hybrid nanorods. *Chem. Comm.* 51: 15039–15042.

119. Kishimoto, Y. and J. Abe. 2009. A fast photochromic molecule that colors only under UV light. *J. Am. Chem. Soc.* 131: 4227–4229.

120. Picot, O. T., M. Dai, D. J. Broer, T. Peijs, and C. W. M. Bastiaansen. 2013. New approach toward reflective films and fibers using cholesteric liquid-crystal coatings. *ACS Appl. Mater. Interfaces* 5: 7117–7121.

121. Wang, J., A. Jákli, and West, J. L. 2015. Airbrush formation of liquid crystal/polymer fibers. *Chemphyschem.* 16: 1839–1841.

122. Vikova, M. 2011. Photochromic textiles. *PhD Thesis.* Heriot-Watt University, Edinburgh.

3 Chiral Nematic Liquid Crystalline Sensors Containing Responsive Dopants

Pascal Cachelin and Cees W. M. Bastiaansen

CONTENTS

3.1 Introduction 63
3.2 Chemosensors 67
    3.2.1 Reactive Cholesterol Derivatives within Cholesteric N* LCs 68
    3.2.2 Responsive Chiral Dopants in Induced N* LCs 70
    3.2.3 Nonchiral Dopants within a N* LC 73
3.3 UV Sensors 74
    3.3.1 UV Sensing via Photoisomerization 75
    3.3.2 UV Sensing via Photoinitiation of a Chemical Reaction 76
    3.3.3 UV Sensing via Photoracemization of a Chiral Dopant 78
3.4 Conclusion 79
References 80

3.1 INTRODUCTION

In his 1990 book, Peter Collings refers to liquid crystals (LCs) as "Nature's delicate phase of matter."[1] This is perhaps most apparent when looking at the use of LCs as sensors. Due to the inherent fragility of the phase, LCs make ideal candidates for sensors, as small disruptions to the localized order are propagated and amplified throughout the bulk material. This responsiveness, coupled with the attractive optical properties of liquid crystalline materials, has been the focus of much of the investigation into the applications of the liquid crystalline phase.

Although LCs are best known for their electro-optical properties, particularly as display materials, investigations into their use for other applications have been ongoing for quite some time. In this chapter, we will be focusing particularly on the sensing applications of chiral nematic (N*) LCs,* although many other applications

---

* Although the terms "cholesteric" and "chiral nematic" are used interchangeably, in this work the term "cholesteric" will refer exclusively to N* LCs formed from cholesterol derivatives, whereas N* LCs formed by the inclusion of a chiral dopant within a nematic LC will be referred to as "induced N* LCs." These two terms encapsulate the two methods of forming a N* LC: either by using a chiral mesogen such as cholesterol myristate or by incorporating a chiral, nonmesogenic molecule into a nematic host.

have been investigated.[2] As the name suggests, N* LCs can rightly be considered an extension of the nematic mesophase. Whereas in the nematic mesophase, the calamitic mesogens align with a common preferred axis (the director), in the N* mesophase, the director rotates through space, causing the mesogens to form a helical superstructure. Throughout a N* helix, there exist regions that share a common director, separated by a half twist of the helix. A distance of two half twists is termed as the pitch (p), which corresponds to the distance required for a full rotation of the director, as shown in Figure 3.1.

The orientation of these helices will be random unless a layer of common alignment is used as a template. This is usually done through the use of an ordered substrate, with a common direction of alignment supplied through methods such as rubbing to induce anisotropy in a polymeric substrate. Such substrates provide a common director for the mesogens to align with, and therefore give order to the bulk of the material. When this happens, we gain common alignment of the N* LC helices, which then begin to show some unusual optical properties.

Regularly repeated layers of molecules sharing an orientation can undergo Bragg reflection of a beam with wavelength (λ) equal to $2d \sin \theta$. In this case, the $2d$ in the equation is dependent on both the pitch and the mean refractive index ($\bar{n}$) of the LC. This is shown below:

$$\lambda = \bar{n}p. \tag{3.1}$$

The mean refractive index in turn is defined in terms of the ordinary (n_o) and extraordinary (n_e) refractive indices of the LC.

$$\bar{n} = \frac{n_e + 2n_o}{3}. \tag{3.2}$$

If the pitch of the N* LC is in the range of 350–600 nm, then the wavelength of the reflected beam will be in the visible region of the electromagnetic spectrum, and the films will adopt a brightly colored hue due to the reflection of a narrow band

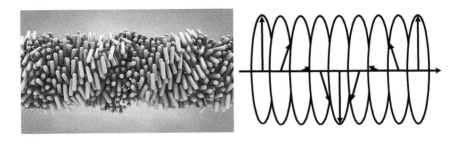

FIGURE 3.1 (See color insert.) An illustration of the N* mesophase. On the left is a computer-generated image showing the alignment of individual molecules, while the image on the right shows the helical progression in the alignment of the director. (Image adapted with permission of Yan Liang/BeautifulChemistry.net.)

of wavelengths, as given by the difference between the ordinary and extraordinary refractive indices.

$$\Delta\lambda = (n_e - n_o)p. \tag{3.3}$$

As a N* helix has a given handedness, only light of polarization matching the handedness of the helix is reflected. This means that a (P)-helix will only reflect right-handed circularly polarized light, while an (M)-helix will only reflect left-handed circularly polarized light.

This unique behavior has been the focus of much of the work into N* LCs. Applications include color filters,[3] polarizers,[4] reflective displays,[5,6] and thermochromic thermometers and paints.[7,8] These are the subject of a recent excellent review by David Coates.[9]

As mentioned previously, there are two main ways to create a N* LC. The first is to use a chiral mesogen which will naturally adopt the helical structure when placed on an ordered substrate. The second is to incorporate a chiral dopant molecule into a nematic LC host. The chiral molecule then acts on the nematic host to create the rotation in the director that gives rise to the N* mesophase. The efficiency of the molecule in inducing a helix is termed as the helical twisting power (HTP) (β) and arises from the geometry of the dopant in question.

Additionally, the optical purity, as measured by the enantiomeric excess (e_e), and the doping concentration (by weight, c_w) have been shown to impact the pitch of the helix created. These are expressed in Equation 3.4:

$$p = (\beta e_e c_w)^{-1}. \tag{3.4}$$

Globally, sensors are a growing field with increasing number of applications in a diverse range of fields. N* LC sensors have to be released into a competitive marketplace which is dominated by several already-mature technologies. In order to be relevant within this marketplace, this sensing platform needs to have significant advantages over the existing technologies in order to compete. One such advantage comes from the fact that N* LC sensors are readily solution processable: Examples have been developed which are made by spincoating,[10] inkjet printing,[11] polymer encapsulation,[12] and spray-coating.[13] These methods represent a highly affordable, high-throughput method of producing devices that does not require expensive production equipment or clean-room facilities. They also integrate well with current manufacturing techniques for a number of products, allowing for the possibility of creating so-called smart materials that respond to environmental conditions. These low-cost methods complement the low cost of the LC materials to produce very cost-efficient disposable sensors.

While sensors utilizing many different mesophases have been developed (notably the work of Ichimura and Abbot on optical command surfaces),[14] N* LC sensors are uniquely placed in that a small change in the structure of the mesophase results in a clear colorimetric signal, allowing for sensors that can be visually assayed even in the absence of access to analytical facilities or electrical power. Many such sensors have been developed. These can be broadly divided into two categories: The first of these relies on the thermally,[8,15] mechanically,[12,16] or chemically induced swelling or shrinking of the N* LC. The second category is the focus of this chapter: Changes in pitch induced by a reaction between a responsive component* of a N* LC and an analyte, as illustrated in Figure 3.2. Both of these topics have been the subject of recent reviews, such as those by White in 2010[17] or Mulder in 2014.[18]

Of the former category, one common method of creating chemosensors is to monitor the change in pitch caused by the swelling or shrinkage of the bulk LC following the absorption of volatile organic compounds (VOCs). Such systems were first expounded by Fergason in 1964,[15] and then subsequently refined, including recent steps such as the incorporation of quartz crystal microbalances to distinguish between VOCs,[19] creating hydrogen-bonded systems to induce preferential absorption,[20] or creating polymer–LC composites in order to monitor environmental conditions such as temperature[21] or humidity.[22] Such systems tend to suffer from one common problem, however. As swelling can be induced by a wide range of substances and conditions, it is difficult to selectively detect a single analyte under noncontrolled conditions. In order to circumvent this, researchers have lately been turning their attention toward the second type of sensor: those based on responsive components within N* LCs.

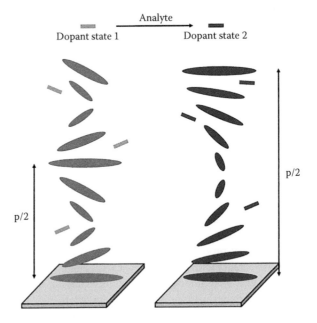

FIGURE 3.2 A schematic showing the change in pitch of a N* LC induced by a reaction between a responsive chiral dopant and an analyte.

---

* Components which undergo a chemical reaction in response to an analyte are termed as responsive components throughout this work. This is to avoid confusion with already existing terms such as reactive mesogens.

By incorporating a responsive chiral dopant that undergoes a chemical reaction in the presence of a desired analyte into a nematic LC host, it is possible to form a N* LC that changes its optical properties in response to that reaction occurring (Figure 3.2). This has the advantage of allowing for much more selective monitoring of analytes: Careful selection of the reactive functional groups present within the chiral dopant allows for specific functionalities present in an analyte to be targeted. It is worth noting that this reactive property does not prevent the inclusion of VOCs into the nematic host. Thus, these sensors are not suitable for use in environments with significant ambient levels of VOCs, as the swelling associated with the absorption of VOCs could easily mask any changes in pitch due to the presence of the targeted analyte.

One further benefit of sensors based on the use of a responsive chiral dopant within a nematic matrix is that it allows for the possibility of creating both sensors and dosimeters, depending on the nature of the interaction between the analyte and the dopant. If the reaction occurring is reversible, then the device acts as a real-time sensor: It actively monitors the current concentration of the analyte. If the reaction is irreversible, then the device acts instead as a time-integrating sensor: The total exposure over a given time frame can be determined. Both of these can be advantageous, depending on the envisioned application, and both will be discussed in this work.

It is also important to acknowledge the shortcomings associated with any sensing platform. One key limitation of these sensors is the requirement for dopants that are both highly reactive while still possessing a sufficiently high HTP in order to induce a N* phase. For visual monitoring, it is also important that large changes in HTP occur as a result of the interaction between the dopant and the analyte. One method of countering this shortcoming is to use molecules known to possess high HTPs as a template to functionalize.

Sensors of this type that have been developed can further be split into two categories: those that have been used to sense chemical analytes (Section 3.2) and those that are triggered by UV light (Section 3.3). These will be discussed in turn below.

3.2 CHEMOSENSORS

The ability to sense the presence or the absence of certain compounds is a highly desirable one. From common household systems such as carbon monoxide detectors to more complex systems for diagnosing diseases or determining the authenticity of works of art, the ability to detect the presence of certain molecules has found many applications. These vary in complexity from simple strips of paper impregnated with chemicals (such as litmus paper) to highly complex and sensitive machines such as a gas chromatograph.

In general, chemosensors refer to devices that are designed to detect a single molecule or group of compounds, compared to analytical machines that are designed to be able to detect a wide range of compounds, such as a mass spectrometer. A device acts as a chemosensor when it selectively produces a signal in the presence of a given analyte. In the case of N* LC chemosensors, this signal comes in the form of a change in the wavelength of maximum reflection (λ_{max}) of the N* LC.

There are three detection methods in N* LCs that have been examined as potentially producing usable chemosensors. The first of these is based on the use of a responsive mesogen (such as a functionalized cholesterol derivative) within a cholesteric LC film. The second is the use of a responsive chiral dopant within an induced N* LC, and the third on the use of responsive achiral dopants in N* LCs.

3.2.1 REACTIVE CHOLESTEROL DERIVATIVES WITHIN CHOLESTERIC N* LCs

Reactive cholesterol derivatives were the first responsive chiral components investigated for sensing applications. An example of this was published by Shinkai et al. in 1990.[23] They had previously developed a crown-ether functionalized cholesterol derivative in order to study selective ion transport (Figure 3.3).[24] In the latter work, they studied the impact of incorporating this into a blend of mesogenic cholesterol derivatives. They found that when exposed to alkaline metal salts, the change in the pitch varied with both the identity of the metal and the counterion. The changes as a result of the anion were attributed to the size of the anion (Figure 3.3). In the case of the metal ions, this relationship did not hold, with K^+ inducing a substantially larger change than Cs^+, despite being significantly smaller (Figure 3.3). Although this effect was not explained, the fact that changing the crown-ether moiety from 15-crown-5 to 18-crown-6 resulted in a significantly lower λ_{max} value for K^+, Li^+, and Na^+, while that of Cs^+ increased, suggesting that the strength of interaction between the crown-ether moiety and the metal cation is responsible for the change in reflection, as opposed to the change being entirely attributable to swelling.

The latter work was built on this by utilizing a second effect of chirally functionalized crown-ethers: their ability to discriminate between enantiomers of ammonium compounds.[25] Nishi et al. utilized the previously described crown-ether-functionalized cholesterol derivatives in order to measure the difference in pitch

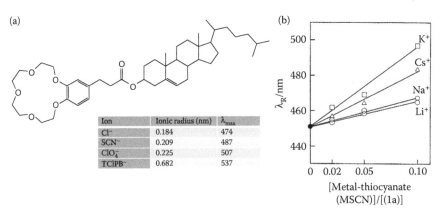

FIGURE 3.3 (a) Crown-ether-functionalized cholesterol derivative 1a.[23] Table shows the Ionic radii and wavelengths of maximum reflection for N* LC films containing crown-ether-functionalized cholesterol derivatives when exposed to various anions. $TClPB^-$ = tetrakis(4-chlorophenyl) borate. (b) A plot of reflection maxima (λ_{max}) against metal thiocyanate concentration for a variety of alkali metals. (S. Shinkai et al., *J. Chem. Soc. Chem. Commun.*, 1990, 1, 303. Reproduced by permission of The Royal Society of Chemistry.)

when cholesteric blends were exposed to a variety of optically pure ammonium salts.[26] They found that D-isomers usually induced significant increases in pitch, while L-isomers induced smaller decreases in pitch. When the cholesterol moiety on the reactive dopant was replaced with a nonchiral alternative, most of the chiral analytes caused no change in the pitch, while others caused very small changes ($\Delta\lambda_{max} < 10$ nm). This confirmed that a reaction of an analyte with a chiral dopant within a LC matrix could induce a change in pitch.

In 1995, Tokuhisa et al. found that using a chiral nematic–polymer composite films containing crown-ether-functionalized cholesterol derivatives lowered the activation energy for ion conduction compared to systems without the ordered arrangement of crown-ether moieties.[27] As might be expected, at higher temperatures where the mesophase becomes increasingly disordered, conductivity decreased. By incorporating an *azo*-moiety (*azo* = R–N=N–R) within the cholesterol derivative, it was further possible to create a photo-switchable ion conductor, as the mesophase moved between ordered and disordered states, although the switching times were significant. Later investigation by Kado showed that the use of 15-crown-5-ether-functionalized cholesterol benzoate (15-crown-5-CB) selectively interacts with K^+ ions.[28] Films containing 15-crown-5-CB exposed to 0.1 M solutions of Li^+, Na^+, and K^+ exhibited a change in both the peak intensity and the value of λ_{max} ($\Delta\lambda_{max} = 18$ nm) only when exposed to solutions of K^+. The dose response was found to be approximately linear in the concentration range of 0–0.1 M, with concentrations above 0.1 M resulting in no further change in the reflection maximum. Larger shifts were possible if the ratio of K^+ to 15-crown-5-CB was >1, although this was accompanied by higher variability in the value of λ_{max}. This was built on the work by Kimura et al., who had first noted the ability of cholesteric LCs containing crown-ether-functionalized cholesterol derivatives to discriminate between Na^+ and K^+.[29]

One later development of this technology was to create more targeted systems. James et al. used boronic acid-functionalized cholesterol derivatives in order to discriminate between different saccharide configurations.[30] James highlighted a number of different complexes that could be formed between boronic acids and saccharides. They found that the dihedral angle between the phenyl moieties of the boronic acids in a 2:1 complex with a saccharide correlated well with the observed shift in λ_{max}. This change is likely due to a change in HTP, although this was not investigated.

One recent example is the use of trifluoroacetyl (TFA)-functionalized cholesterol derivatives by Kirchner et al. to selectively detect amine vapors.[31] This built on work in developing chromogenic and fluorogenic biogenic amine sensors using the TFA moiety, while trying to avoid the limited lifetime associated with the photoinduced decomposition of chromogenic and fluorogenic molecules.[32] As the color associated with N* LCs is structural, as opposed to that resulting from electronic excitation of molecules, there is greater inherent stability. On exposure to amine vapors, films containing TFA-functionalized cholesterol derivatives along with nonreactive cholesterols showed a linear dose-dependent change in the value of λ_{max}, with a shift of 40 nm observed for 1600 ppm of 1-butylamine (Figure 3.4).

While the systems developed in this way showed good selectivity toward amines, with no signal observed when exposed to ketones, alcohols, or acetates, the systems

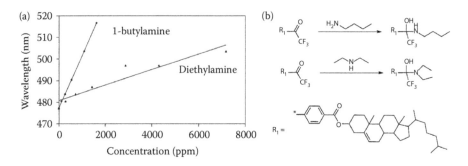

FIGURE 3.4 (a) Graph showing the change in λ_{max} between varying concentrations of n-butylamine and diethylamine for the TFA-functionalized cholesterics developed by Kirchner et al.[31] (b) The formation of the hemiaminal products of the reaction between the TFA-functionalized cholesterol and the amines investigated. (N. Kirchner et al., *Chem. Commun. (Camb).*, 2006, 1512–1514. Reproduced by permission of The Royal Society of Chemistry.)

are not able to distinguish between amines which limit their utility in real-world applications. This problem is a common one to systems based on reactive dopants, although N* LC polymer systems based on swelling and shrinkage that are able to discriminate between primary alcohols have been developed.[33]

Systems based on functionalized cholesterol derivatives have both associated advantages and disadvantages. One strong benefit is a known compatibility with a mesogenic system. This avoids the potentially time-consuming process of screening LC hosts. There is also the additional benefit that HTP considerations are not as important: The other components of a cholesteric liquid crystalline mix will ensure a visible reflection band. This latter property is also the basis of a significant disadvantage associated with functionalized cholesterol derivatives: Small changes to the reactive component can be masked by the other contributing chiral mesogens. This contrasts to an induced N* LC, where the nematic host amplifies small changes within the responsive chiral dopant.

3.2.2 Responsive Chiral Dopants in Induced N* LCs

As mentioned in the introduction to this chapter, induced N* LCs are exciting potential sensor materials. By incorporating a responsive chiral dopant, it is possible to create devices with high sensitivity; though the dopant is usually a small component in the overall liquid crystalline mixture yet it is solely responsible for inducing the helical twist of the N* mesophase. It therefore follows that a chemical reaction involving the dopant is a highly efficient way of inducing an optical change within a N* LC thin film.

This latter area of study is very recent, with the first developed systems dating from 2010.[34] Han et al. developed two sensors, a CO_2 sensor that showed real-time monitoring and an O_2 sensor that acted as a time-integrating sensor. These two different behaviors are rooted in the nature of the reaction between the chiral dopant and the analyte. If the reaction is reversible, or if the analyte forms a weakly held complex with the dopant as opposed to forming new chemical bonds, then the sensor will regenerate over time, although device hysteresis is a significant issue. If on the other hand the reaction is irreversible, then the device will record the total dose of the analyte, acting as a time-integrating sensor, or dosimeter. Both of these behaviors are desirable, depending on context.

In the case of the CO_2 sensor, a diamine was used as the responsive chiral dopant of the mixture, which would reversibly react with CO_2, forming a carbamate intermediate. The diamine chosen was found to have a very low HTP in order to induce visible reflection bands. In order to avoid this issue, the diamine was complexed with a tetraaryldioxolanediol (TADDOL) derivative, which is known to possess very high HTPs (Figure 3.5).[35] When the films containing 1.6% of the diamine–TADDOL complex were exposed to an atmosphere of pure CO_2, a rapid (<10 s) shift from an original red state to a blue state was observed. This was attributed to the carbamate-TADDOL complex having a higher HTP than the unreacted dopant ($202 \ \mu m^{-1}$ for the carbamate–TADDOL complex, compared to $157 \ \mu m^{-1}$ for the diamine–TADDOL complex). The reverse reaction was considerably slower, with an estimated 2.5 h required to reach 50% decomplexation. Studies at lower concentrations of CO_2 found that at concentrations below 10% CO_2, complete conversion was no longer possible, and the time taken to reach a stable conversion percentage was significantly longer. Due to the difference in rate between the forward and reverse reactions at low concentrations, the devices started displaying semidosimetric behavior, an undesirable characteristic in a real-time sensor.

In the case of the O_2 sensor, a thiolated bis-2-naphthol (BINOL) derivative (BINSH) was used as the active chiral dopant (Figure 3.6). As before, the solubility and HTP of the dopant were too low in order to create a visible reflection band (BINSH $\beta = 13 \ \mu m^{-1}$ in E7). In this case, a related codopant was used in order to create a visible reflection band. In the presence of a small amount of an amine catalyst, the BINSH underwent an irreversible reaction when exposed to oxygen, with no change in λ_{max} observed when exposed to N_2, or when exposed to air in the absence of the triethylamine catalyst. As the disulfide product has a higher HTP than the BINSH dopant (disulfide $\beta = 65 \ \mu m^{-1}$), a blue shift would be expected. In this case

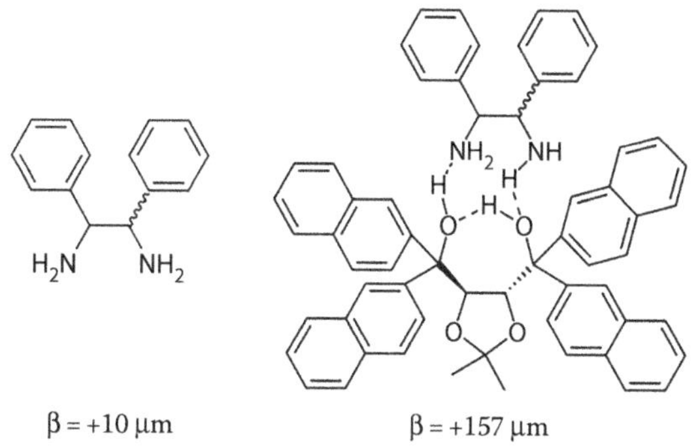

$\beta = +10 \ \mu m$ $\beta = +157 \ \mu m$

FIGURE 3.5 The structure and helical twisting power of 1,2-diphenylethane-1,2-diamine before and after complexation with TADDOL.

FIGURE 3.6   The structure of BINSH chiral dopant before and after the reaction with O_2.

instead a red shift was observed, which is attributed to the substantially lower solubility of the disulfide product in the nematic host, resulting in precipitation effectively lowering the concentration of the dopant within the N* LC.

A similar behavior was observed in the case of a humidity sensor developed by Saha et al.[36] This sensor was based on the creation of a BINOL dimer, which was expected to have a significantly higher HTP than the monomeric unit, which has been found to have a low HTP in nematic hosts such as E7.[37] This expectation was based on a previous work showing an increase in HTP of bridged biaryls compared to their open-chain counterparts.[38] This increase is attributed by Eelkema and Feringa to a change in the dihedral angle formed between the two naphthyl units.[39] BINOL derivatives have the ability to move from transoid to cisoid forms (Figure 3.7), depending on the groups present on the 1-naphthyl site has also been identified as the reason why BINOL derivatives of either chirality can form both right- and left-handed helices.[38,39]

When exposed to water vapor, the BINOL dimer undergoes an irreversible reaction which returns it to the monomeric state, which due to lowered HTP and solubility in the nematic host resulting in a notable increase in the wavelength of the reflection maximum (Figure 3.8).

An irreversible reaction between a chiral dopant and an analyte was also the basis of a recent work by Cachelin et al. looking at acetone sensors.[10] In this work, a complex was formed between an achiral hydrazine and TADDOL, creating a complex that selectively reacted with acetone (Figure 3.9). The dosimetric behavior of the sensor was analyzed in this case, with a linear dose response observed over the operational range of the sensor. It is worth noting the small change in λ_{max} observed

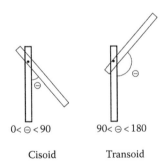

FIGURE 3.7   Illustration of the transoid and cisoid BINOL forms, which are dependent on the dihedral angle θ.

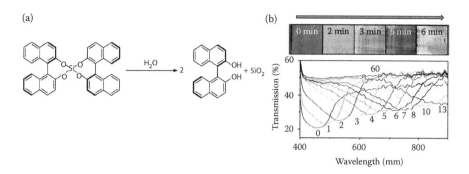

FIGURE 3.8   (a) The reaction between the humidity-sensitive BINOL dimer and H_2O. (b) The change in wavelength of the reflection maximum as observed as a function of time exposed to air of 75% relative humidity. (A. Saha et al., *Chem. Commun.*, 2012, 48, 4579. Reproduced by permission of The Royal Society of Chemistry.)

in the case of this acetone sensor. This was attributed to a very small change in HTP as a result of the reaction with the analyte ($\Delta\beta = 2.08\ \mu m^{-1}$). Nevertheless, the fact that a dosimetric behavior was observed even when the changes involved in the molecular geometry of the dopant were very small indicates that this is a very sensitive technique for the monitoring of volatile analytes.

3.2.3   NONCHIRAL DOPANTS WITHIN A N* LC

One final method of creating chemosensors is to incorporate an achiral dopant within a N* LC in order to change the liquid crystalline matrix. Currently there is only one such example, based on the inclusion of magnetite nanoparticles with a cholesteric N* LC host. It has previously been shown that the incorporation of magnetite nanoparticles within a LC sample results in a change in their electrical and magnetic properties. As magnetite is similarly known to show good selectivity toward carbon monoxide (CO); Aksimentyeva et al. investigated the possibility of creating a sensor from a N* LC doped with magnetite nanoparticles (Figure 3.10). On exposure to

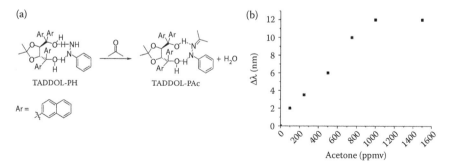

FIGURE 3.9   (a) The reaction scheme for the chiral dopant and acetone. (b) Change in λ_{max} plotted against acetone concentration. The plateau at concentrations above 1000 ppm was attributed to the saturation of the sensor. (P. Cachelin et al.: *Adv. Opt. Mater.*, 592–596. 2016. Copyright Wiley-VCH Verlag GmbH & Co. KGaA. Reproduced with permission.)

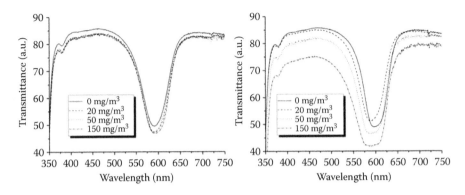

FIGURE 3.10 Reflection bands of cholesteric LCs both with (right) and without (left) magnetite nanoparticles on being exposed to different concentrations of CO. (Image adapted from O. Aksimentyeva et al., *Mol. Cryst. Liq. Cryst.*, 2014, 589, 83–89, with permission from Taylor & Francis.)

low concentrations of CO, it was found that cholesteric solutions doped with 0.3%–0.67% of magnetite nanoparticles underwent concentration-dependent peak broadening, with an accompanying small shift in λ_{max}.[40] This was attributed to competitive adsorption between CO and the LC host on the magnetite surface. Samples that did not include the magnetite nanoparticles did not display this behavior. Additionally, the optical density of the films changed in response to CO concentration, although this was not expanded upon by the authors. Of particular note in this work is that both the low concentration of the dopant and the very fast reaction times (<5 min), are highly dependent on the kinetics of the reaction under investigation.

3.3 UV SENSORS

It should come as no surprise that much of the early attention in the field of responsive liquid crystalline materials was based on their optical switching behavior. The ability to optically address or pattern displays was, and remains, an attractive alternative to electrical addressing. Much of the work in this section is therefore focused on this type of behavior, with relatively little attention being paid to these materials as sensors. Nonetheless, these materials do act as sensors; they respond to a stimulus by changing some property of themselves. Unfortunately, sensing properties, such as the sensitivity or dose dependence, are less explored in these materials.

Using N* LCs as UV sensors has a long history, with the first mention dating from a 1965 technical report by Fergason et al., where a mention is made of the existence of UV-sensitive cholesteric LCs that respond by changing their color.[41] This was not the focus of their investigation, however, and no further details were given. Since that time, UV-responsive N* LCs have remained an area of active investigation.

As well as chemically interacting with the dopant, as described in Section 3.2, it is possible to induce a chemical change by nonchemical means, such as UV radiation. Photoisomerization, photoinitiated reactions, and photoracemization are methods that can induce a chemical change within the dopant molecule, which can then be propagated to the nematic matrix.

The first of these, photoisomerization, is the basis of the optical command surfaces first developed by Ichimura et al. in 1988.[42] These are covered elsewhere within this book. There have also been applications of azodyes as free dopants within N* LCs, as discussed in Section 3.3.1.

Photoinitiation is the basis of a group of related sensors which utilize the photoinitiated transformation of the sterol-based D provitamins such as ergosterol and 7-dehydrocholesterol. These are based on change in HTP between the provitamin and vitamin forms, and thus are analogous to the reactive dopants discussed in Section 3.2.2.

Another method of sensing UV radiation is through the racemization of a photoracemizable dopant. The effect of this is to change the enantiomeric excess of the dopant, which should result in a red shift in the associated reflection band. This is discussed later in this chapter.

3.3.1 UV Sensing via Photoisomerization

Photoisomerization is the process of conversion between the *cis* and *trans* isomers across a double bond, depending on the wavelength of illumination. Azodyes are one such example of the compounds that undergo this isomerization. By incorporating azobenzene within a cholesteric host, Sackmann showed that it was possible to reversibly change the value of λ_{max} by selective illumination.[43] On exposure to light with a wavelength of 420 nm, a solution containing only the *cis* azobenzene showed a blue shift from $\lambda_{max} = 610$ nm to $\lambda_{max} = 560$ nm (Figure 3.11). On exposure to light with a wavelength of 420 nm, a solution containing only the *cis* azobenzene showed a blue shift from 610 to 560 nm. Similarly, a red shift was observed when exposing a solution containing only the *trans* isomer to light with a wavelength of 313 nm. As the photoisomerization of azobenes is a rapid process, the change in λ_{max} was similarly rapid, although the kinetics of the change was not established.

A second example of such a photoisomerization can be seen in the work by Yarmolenko et al. in 1994.[44] Here again a photosensitive chiral compound was exposed to UV light, which underwent a corresponding shift in HTP (Figure 3.12).

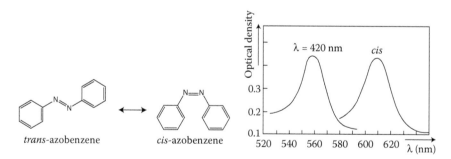

FIGURE 3.11   (a) Chemical structure of *cis* and *trans* azobenzene. (b) Change in λ_{max} between the trans and cis azobenzene within a cholesteric host. (Reprinted with permission from E. Sackmann, *J. Am. Chem. Soc.*, 7088–7090. Copyright 1971 American Chemical Society.)

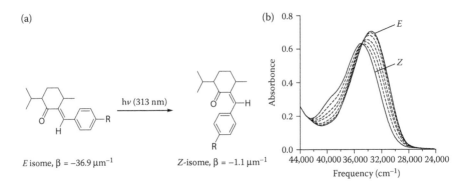

FIGURE 3.12 (a) Structure of the isomers. β values are in 5CB. (b) Absorption spectra of the E and Z isomers of 2-([1,1′-biphenyl]-4-ylmethylene)-6-isopropyl-3-methylcyclohexan-1-one. The dash lines are the spectra resulting from progressive irradiation of the E isomer. (Image adapted from S. N. Yarmolenko et al., *Liq. Cryst.*, 1994, 16, 877–882.)

Of particular interest are the significantly different values of β associated with the *E* and *Z* isomers. This was attributed by the authors to the *pseudo*-calamitic shape of the *E* isomer allowing for a stronger interaction with the nematic host, as opposed to the *pseudo*-spherical *Z* isomer. This was supported by infrared (IR) spectroscopy studies indicating a significant nonplanar character associated with the carbonyl moiety of the *Z* isomer.

This principle was extended by Feringa et al. in 1995 to produce a device that switched the mesophase on UV illumination.[45] To do this, they used a molecule that they had been previously studied as a chiroptical molecular switch.[46] Chiroptical switching is a form of molecular switching where photochemical switching occurs between enantiomers, often those exhibiting planar chirality such as hindered alkenes, which can readily undergo isomerization.[45] As the two stable states of the molecular switch used exhibit opposite helicity, it was possible to reach an intermediate state that did not contain an excess of either dopant. When the two forms were balanced, the LC film adopted a compensated nematic mesophase, which could be monitored by polarized optical microscopy. This allowed for an effective three-position photochemical switch, due to the differences in the polarization of light transmitted by the device in the three different states.

One attractive feature of photoisomerizable UV systems is the existence of two stable states, allowing them effectively to act as photo-addressable switches. This is less beneficial when considering their use as sensors, as natural UV sources such as solar irradiation output a wide range of wavelengths and can trigger both the forward and reverse reactions. In such cases, a UV-induced irreversible photochemical reaction is more desirable, as it allows for both dosimetric sensing as well as not suffering from photoinduced reverse reactions.

3.3.2 UV Sensing via Photoinitiation of a Chemical Reaction

Another method of sensing UV light within a N* LC is to rely on the photoinitiation of a chemical reaction involving the dopant. One early work in this area by Haas

FIGURE 3.13   The effect of sequential exposure on a mixture of cholesterol iodide and cholesterol nonanoate. Light = red, dark = clear. (Image reproduced from W. Haas et al., *Mol. Cryst.*, 1969, 7, 371–379 with permission from Taylor & Francis.)

et al. drew on the reference from Fergason et al. for inspiration, and exposed a variety of cholesterol derivatives to UV light of 300 nm wavelength to measure their response.[47] They found that mixtures of cholesteryl halides and aliphatic cholesteryl esters would undergo a change from clear to colored, with a dosage-dependent red shift observed (Figure 3.13). This was attributed to the breaking of the photolabile cholesteryl-iodide bond, and the subsequent formation of a new compound. Although the structure of this new compound was not determined, this assertion was supported by the fact that a reduction in the concentration of cholesteryl iodide without UV exposure resulted in a blue shift.

Much of the later work for these dopants rested on the transformation of certain sterols on exposure to light. One well-known example of this is the transformation of provitamin D_2 to vitamin D_2 on exposure to UV light, which has important ramifications for human health (Figure 3.14).

The two-step conversion of provitamin D_3 into previtamin D_3, followed by the isomerization to tachysterol was the basis of a personal UV dosimeter proposed by Terenetskaya and Gvozdovsky.[48] They found that UV exposure resulted in a significant decrease in the apparent pitch of a N* LC, as measured by the number of lines present within a Grandjean-Cano wedge cell. These lines result from the change in distance between the surfaces of the cells changing the number of N* helices that can fit within the cell, with dark lines appearing regularly as a result. By measuring

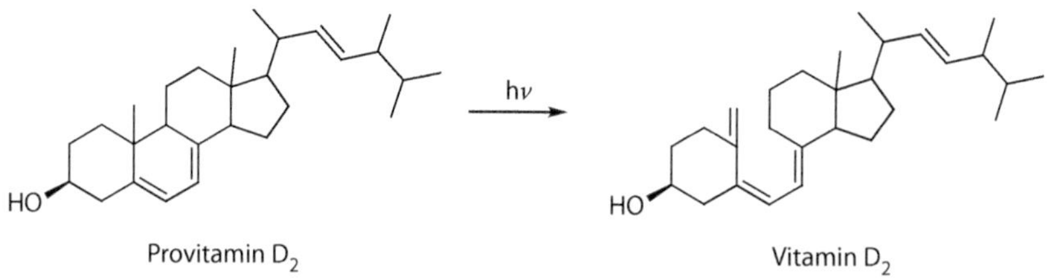

HO — Provitamin D_2 $h\nu$ HO — Vitamin D_2

FIGURE 3.14   The UV-initiated transformation of provitamin D_2 to vitamin D_2.

the distance between lines and the angle of the cell (α) the apparent pitch can be determined, using Equation 3.5.

$$P = 2S \tan \alpha. \tag{3.5}$$

This apparent change in pitch was attributed to the transformation of provitamin D_3 in tachysterol, although the specific HTP of tachysterol was not measured. Although the initial devices had pitches that were significantly larger than those that would yield visible reflection bands, further work by the group succeeded in creating devices with visible reflection bands, and the impact of LC composition and UV dose on the pitch was closely investigated.[49,50]

As Petriashvili et al. noted, one major drawback of these systems is that it is not possible to extract UV dose, due to the rapid nature of the chemical reaction behind the change in pitch.[51] Their response was the creation of a N* LC containing a dye that absorbed UV photons and emitted within the visible spectrum.[51] Their response was the creation of a chiral nematic LC containing a dye that absorbed UV photons and emitted within the visible spectrum. This system, when coupled with a photodiode, could then quantitatively measure the UV exposure of the system. While this avoids the use of expensive UV photodiodes in favor of cheaper visible light photodiodes, the use of a fluorogenic compound does place the system at risk from photo-decomposition, limiting the device's lifetime significantly.

3.3.3 UV Sensing via Photoracemization of a Chiral Dopant

One method of UV sensing that has been little explored is the racemization of a chiral dopant. This works as a highly effective sensing method and is based on the enantiomeric excess of the dopant. The enantiomeric excess of a dopant is defined in Equation 3.6.

$$ee = \frac{[D_L] - [D_R]}{[D_L] + [D_R]} \tag{3.6}$$

For an enantiomerically pure dopant, $ee = 1$. Likewise, a completely racemic dopant will give a value of $ee = 0$. If the ee of a dopant reaches zero, the pitch of the cholesteric becomes infinity large, thereby modeling the expected N*–nematic transition. Changes below this threshold should show a progressive red shift. This was demonstrated by van Delden and Feringa in 2001, where the reaction between a chiral compound of unknown ee and a reactive LC resulted in the formation of a N* mesophase where the value of λ_{max} could be used to determine the ee of the compound.[52] They did however note that the value of $\Delta\lambda_{max}$ obtained depended both on the concentration as well as the enantiomeric purity of the sample.[52]

This process was first observed by Mioskowski et al. in 1976.[53] Using dissymmetric photolabile compounds, they found that the phase change was clearly visible using polarized optical microscopy. LC droplets of K15 containing a photolabile chiral dopant showed significant changes in pitch after 3 h of irradiation at 15 kW/m², with a

N*–nematic transition being observed after 3 days. They also found that the extent of polarization depended on the relative orientation of the N* helix and light source, with a parallel orientation resulting in significantly faster racemization compared to a perpendicular one, with the fastest racemization observed when the sample was exposed as an isotropic droplet.

A second example of this sensing mechanism has been recently developed by Cachelin et al.[54] This sensor used the photoracemization of (R,R')-bis-2-napthol (BINOL) as the host material. This material had previously been studied by Zhang et al. as a potential photo-switchable dopant, but the dopant was shown not to undergo chiral enrichment when illuminated with circularly polarized light.[55] By comparing the rate of racemization as measured by circular dichroism (CD) spectroscopy with that obtained through thin films of N* LCs containing the photoracemizable dopant, Cachelin et al. were able to show that the change in pitch can be modeled by Equation 3.7, where e_f and e_i are the enantiomeric excess of the final and initial states, respectively.

$$\Delta p = \frac{e_f - e_i}{e_f e_i \beta c_w}. \tag{3.7}$$

This showed that sensors utilizing racemization as the basis of detection are likely to have far higher sensitivities overall then those based on a change in β, such as those mentioned in Section 3.2.2.

Sensors based on photoracemization are a relatively new field, but initial results suggest that it is a highly promising area for future investigations. Another potential area for study is the use of other triggers for racemization, such as chemical substitutions that proceed via an achiral intermediary state. According to the model developed by Cachelin et al., these sensors should exhibit far higher sensitivities than the systems discussed in Section 3.2, and could potentially further boost the performance of N* LC chemosensors.

3.4   CONCLUSION

Overall, we have seen many different approaches to the same principle: The use of a responsive compound within a N* LC in order to detect some analytes, usually by changing the optical properties of the N* LC material. The fact that this method can be applied to such a wide range of analytes speaks strongly in its favor, as does the fact that the output of such systems can usually be determined visually, without the need for complex analytical facilities. We have also seen that the detection limits of such systems are limited, and usually fall within the ppm range, and that detection times tend to be on the scale of minutes to hours as opposed to seconds. The fact that both real-time and dosimetric systems are accessible, depending on the nature of the interaction between the dopant and the analyte, is another point showing the flexibility of this nascent technology. It is important to therefore carefully consider the intended application of these sensors. Their low cost and size, combined with the potential to create dosimetric sensors, are highly desirable for applications such as environmental monitoring and for packages in transit (so-called smart packaging). Likewise, the fact that they produce a clear colorimetric signal without the need to electrical power or access to analytical facilities makes them ideally suited for consumer-based sensing applications, such as personal UV dosimetry. It is important therefore that future applications of this technology focus on exploiting and expanding on these strengths, as opposed to attempting to outcompete current technologies in terms of sensitivity.

REFERENCES

1. P. J. Collings, *Liquid Crystals*, Adam Hilger, Bristol, 1st edn., 1990.
2. Q. Li, Ed., *Liquid Crystals Beyond Displays*, Wiley, Hoboken, New Jersey, 1st edn., 2002.
3. J. Lub, P. van de Witte, C. Doornkamp, J. P. A. Vogels and R. T. Wegh, Stable photo-patterned cholesteric layers made by photoisomerization and subsequent photopolymerization for use as color filters in liquid-crystal displays, *Adv. Mater.*, 2003, 15, 1420–1425.
4. D. J. Broer, J. Lub and G. N. Mol, Wide-band reflective polarizers from cholesteric polymer networks with a pitch gradient, *Nature*, 1995, 378, 467–469.
5. D. K. Yang, J. L. West, L. C. Chien and J. W. Doane, Control of reflectivity and bistability in displays using cholesteric liquid crystals, *J. Appl. Phys.*, 1994, 76, 1331–1333.
6. D. K. Yang, L. C. Chien and J. W. Doane, in Conference Record of the 1991 International Display Research Conference, IEEE, Liquid Crystal Institute, Kent State University, Ohio, pp. 49–52.
7. P. T. Ireland and T. V. Jones, The response time of a surface thermometer employing encapsulated thermochromic liquid crystals, *J. Phys. E.*, 1987, 20, 1195–1199.
8. I. Sage, Thermochromic liquid crystals, *Liq. Cryst.*, 2011, 38, 1551–1561.
9. D. Coates, Development and applications of cholesteric liquid crystals, *Liq. Cryst.*, 2015, 42, 1–13.
10. P. Cachelin, J. P. Green, T. Peijs, M. Heeney and C. W. M. Bastiaansen, Optical acetone vapor sensors based on chiral nematic liquid crystals and reactive chiral dopants, *Adv. Opt. Mater.*, 2016, 4, 592–596.
11. J. E. Stumpel, C. Wouters, N. Herzer, J. Ziegler, D. J. Broer, C. W. M. Bastiaansen and A. P. H. J. Schenning, An optical sensor for volatile amines based on an inkjet-printed, hydrogen-bonded, cholesteric liquid crystalline film, *Adv. Opt. Mater.*, 2014, 2, 459–464.
12. O. T. Picot, M. Dai, E. Billoti, D. J. Broer, T. Peijs and C. W. M. Bastiaansen, A real time optical strain sensor based on a cholesteric liquid crystal network, *RSC Adv.*, 2013, 3, 18794–18798.
13. O. T. Picot, M. Dai, D. J. Broer, T. Peijs and C. W. M. Bastiaansen, New approach toward reflective films and fibers using cholesteric liquid-crystal coatings, *ACS Appl. Mater. Interfaces*, 2013, 5, 7117–7121.
14. R. J. Carlton, J. T. Hunter, D. S. Miller, R. Abbasi, P. C. Mushenheim, L. N. Tan and N. L. Abbott, Chemical and biological sensing using liquid crystals, *Liq. Cryst. Rev.*, 2013, 1, 29–51.
15. J. L. Fergason, Liquid crystals, *Sci. Am.*, 1964, 211, 76–85.
16. P. V. Shibaev, A. Iljin, J. Troisi and K. Reddy, Highly viscous chiral thin films for optical detection of small rotations and submicron displacements, *Appl. Phys. A*, 2013, 114, 339–346.
17. T. J. White, M. E. McConney and T. J. Bunning, Dynamic color in stimuli-responsive cholesteric liquid crystals, *J. Mater. Chem.*, 2010, 20, 9832.

18. D. J. Mulder, A. P. H. J. Schenning and C. W. M. Bastiaansen, Chiral-nematic liquid crystals as one dimensional photonic materials in optical sensors, *J. Mater. Chem. C*, 2014, 2, 6695–6705.
19. A. Mujahid, H. Stathopulos, P. A. Lieberzeit and F. L. Dickert, Solvent vapour detection with cholesteric liquid crystals—Optical and mass-sensitive evaluation of the sensor mechanism, *Sensors*, 2010, 10, 4887–4897.
20. D. J. Broer, C. M. W. Bastiaansen, M. G. Debije and A. P. H. J. Schenning, Functional organic materials based on polymerized liquid-crystal monomers: Supramolecular hydrogen-bonded systems, *Angew. Chem. Int. Ed. Engl.*, 2012, 51, 7102–7109.
21. D. J. D. Davies, A. R. Vaccaro, S. M. Morris, N. Herzer, A. P. H. J. Schenning and C. W. M. Bastiaansen, A printable optical time-temperature integrator based on shape memory in a chiral nematic polymer network, *Adv. Funct. Mater.*, 2013, 23, 2723–2727.
22. J. E. Stumpel, E. R. Gil, A. B. Spoelstra, C. W. M. Bastiaansen, D. J. Broer and A. P. H. J. Schenning, Stimuli-responsive materials based on interpenetrating polymer liquid crystal hydrogels, *Adv. Funct. Mater.*, 2015, 25, 3314–3320.
23. S. Shinkai, T. Nishi, A. Ikeda, T. Matsuda, K. Shimamoto and O. Manabe, Crown-metal interactions in cholesteric liquid crystals, *J. Chem. Soc. Chem. Commun.*, 1990, 1, 303.
24. S. Shinkai, K. Shimamoto, O. Manabe and M. Sisido, Selective ion permation across membranes containing crown ethers with steroid moieties, *Makromol. Chem., Rapid Commun.*, 1989, 10, 361–366.
25. E. B. Kyba, K. Koga, L. R. Sousa, M. G. Sigel and D. J. Cram, Chiral recognition in molecular complexing, *J. Am. Chem. Soc.*, 1973, 97, 2692–2693.
26. T. Nishi, A. Ikeda, T. Matsuda and S. Shinkai, Detection of chirality by colour, *J. Chem. Soc. Chem. Commun.*, 1991, 339.
27. H. Tokuhisa, K. Kimura, M. Yokoyama and S. Shinkai, Ion-conducting behaviour and photoinduced ionic-conductivity switching of composite films containing crowned cholesteric liquid crystals, *J. Chem. Soc. Faraday Trans.*, 1995, 91, 1237.
28. S. Kado, Y. Takeshima, Y. Nakahara and K. Kimura, Potassium-ion-selective sensing based on selective reflection of cholesteric liquid crystal membranes, *J. Incl. Phenom. Macrocycl. Chem.*, 2012, 72, 227–232.
29. K. Kimura, Y. Kawai, S. Oosaki, S. Yajima, Y. Yoshioka and Y. Sakurai, Dependence of ion selectivity on ordered orientation of neutral carriers in ion-sensing membranes based on thermotropic liquid crystals, *Anal. Chem.*, 2002, 74, 5544–5549.
30. T. D. James, K. R. A. S. Sandanayake and S. Shinkai, Determination of the absolute configuration of monosaccharides by a colour change in a chiral cholesteric liquid crystal system, *Angew. Chem., Int. Ed. Engl.*, 1996, 35, 1910–1922.
31. N. Kirchner, L. Zedler, T. G. Mayerhöfer and G. J. Mohr, Functional liquid crystal films selectively recognize amine vapours and simultaneously change their colour, *Chem. Commun. (Camb).*, 2006, 1512–1514.
32. G. J. Mohr, C. Demuth and U. E. Spichiger-Keller, Application of chromogenic and fluorogenic reactands in the optical sensing of dissolved aliphatic amines, *Anal. Chem.*, 1998, 70, 3868–3873.
33. C. K. Chang, C. W. M. Bastiaansen, D. J. Broer and H. L. Kuo, Discrimination of alcohol molecules using hydrogen-bridged cholesteric polymer networks, *Macromolecules*, 2012, 45, 4550–4555.
34. Y. Han, K. Pacheco, C. W. M. Bastiaansen, D. J. Broer and R. P. Sijbesma, Optical monitoring of gases with cholesteric liquid crystals, *J. Am. Chem. Soc.*, 2010, 132, 2961–2967.
35. H. G. Kuball, B. Weiss, A. K. Beck and D. Seebach, TADDOLs with unprecedented helical twisting power in liquid crystals, *Helv. Chim. Acta*, 1997, 80, 2507–2514.
36. A. Saha, Y. Tanaka, Y. Han, C. M. W. Bastiaansen, D. J. Broer and R. P. Sijbesma, Irreversible visual sensing of humidity using a cholesteric liquid crystal, *Chem. Commun.*, 2012, 48, 4579.

37. G. Gottarelli, G. P. Spada, R. Bartsch, G. Solladie and R. Zimmermann, Induction of the cholesteric mesophase in nematic liquid crystals: Correlation between the conformation of open-chain chiral 1,1′-binaphthyls and their twisting powders, *J. Org. Chem.*, 1986, 51, 589–592.

38. G. Gottarelli, M. Hibert, B. Samori, G. Solladi, G. P. Spada and R. Zimmermann, Induction of the cholesteric mesophase in nematic liquid crystals: Mechanism and application to the determination of bridged biaryl configurations, *J. Am. Chem. Soc.*, 1983, 105, 7318–7321.

39. R. Eelkema and B. L. Feringa, Amplification of chirality in liquid crystals, *Org. Biomol. Chem.*, 2006, 4, 3729–3745.

40. O. Aksimentyeva, Z. Mykytyuk, A. Fechan, O. Sushynskyy and B. Tsizh, Cholesteric liquid crystal doped by nanosize magnetite as an active medium of optical gas sensor, *Mol. Cryst. Liq. Cryst.*, 2014, 589, 83–89.

41. C. H. Jones, J. L. Fergason and J. A. Asars, Investigations of large-area display screen using liquid crystals, *RADC Rep.*, 1965, 274.

42. K. Ichimura, T. Seki, A. Hosokit and K. Aoki, Reversible change in alignment mode of nematic liquid crystals regulated photochemically by "command surfaces" modified with an azobenzene monolayer, *Langmuir*, 1988, 1216, 1214–1216.

43. E. Sackmann, Photochemically induced reversible color changes in colesteric liquid crystals, *J. Am. Chem. Soc.*, 1971, 7088–7090.

44. S. N. Yarmolenko, L. A. Kutulya, V. V. Vashchenko and L. V. Chepeleva, Photosensitive chiral dopants with high twisting power, *Liq. Cryst.*, 1994, 16, 877–882.

45. B. L. Feringa, N. P. M. Huck and H. A. van Doren, Chiroptical switching between liquid crystalline phases, *J. Am. Chem. Soc.*, 1995, 117, 9929–9930.

46. B. L. Feringa, W. F. Jager, B. De Lange and E. W. Meijer, Chiroptical molecular switch, *J. Am. Chem. Soc.*, 1991, 113, 5468–5470.

47. W. Haas, J. Adams and J. Wysocki, Interaction between UV radiation and cholesteric liquid crystals, *Mol. Cryst.*, 1969, 7, 371–379.

48. I. Terenetskaya and I. Gvozdovsky, Development of personal UV biodosimeter based on Vitamin D photosynthesis, *Mol. Cryst. Liq. Cryst. Sci. Technol. Sect. A. Mol. Cryst. Liq. Cryst.*, 2001, 368, 551–558.

49. M. Aronishidze, A. Chanishvili, G. Chilaya, G. Petriashvili, S. Tavzarashvili, L. Lisetski, I. Gvozdovskyy and I. Terenetskaya, Color change effect based on provitamin D phototransformation in cholesteric liquid crystalline mixtures, *Mol. Cryst. Liq. Cryst.*, 2004, 420, 47–53.

50. L. N. Lisetski, V. D. Panikarskaya, N. A. Kasyan, L. V. Grishchenko and I. P. Terenetskaya, Bioequivalent UV detectors based on cholesteric liquid crystals: Effects of spectral composition and quantitative account for intensity of UV radiation, *Proc. SPIE*, 2005, 6023, 60230F–60230F–4.

51. G. Petriashvili, A. Chanishvili, G. Chilaya, M. A. Matranga, M. P. De Santo and R. Barberi, Novel UV sensor based on a liquid crystalline mixture containing a photoluminescent dye, *Mol. Cryst. Liq. Cryst.*, 2009, 500, 82–90.

52. R. A. van Delden and B. L. Feringa, Color indicators of molecular chirality based on doped liquid crystals, *Angew. Chem., Int. Ed.*, 2001, 40, 3198–3200.

53. C. Mioskowski, J. Bourguignon, S. Candau and G. Solladie, Photochemically induced cholesteric-nematic transition in liquid crystals, *Chem. Phys. Lett.*, 1976, 38, 456–459.

54. P. Cachelin, H. Khandelwal, T. Peijs, J. Gautrot and C. W. M. Bastiaansen, *Forthcoming Publication*.

55. M. Zhang and G. B. Schuster, Photoracemization of optically active 1,1′-binaphthyl derivatives: Light-initiated conversion of cholesteric to compensated nematic liquid crystals, *J. Phys. Chem.*, 1992, 96, 3063–3067.

4 Cholesteric Liquid Crystalline Polymer Networks as Optical Sensors

Monali Moirangthem and Albert P. H. J. Schenning

CONTENTS

4.1 Introduction 83
4.2 Optical Sensors 85
    4.2.1 pH Sensors 85
    4.2.2 Amino Acid Sensors...86
    4.2.3 Amine Sensors..86
    4.2.4 Humidity Sensors 89
    4.2.5 Alcohol Sensors 90
    4.2.6 Metal Ion Sensors 91
    4.2.7 Temperature Sensors..94
    4.2.8 Strain Sensors 96
4.3 Conclusion 97
References..100

4.1 INTRODUCTION

In the past decade, chiral nematic liquid crystals (LCs) have emerged as an attractive material for the development of stimuli-responsive systems (White et al. 2010; Ge and Yin 2011; Fenzl et al. 2014; Mulder et al. 2014; Stumpel et al. 2014). Due to the periodic alteration of their refractive indices, they act as one-dimensional photonic structures and reflect circularly polarized light of same handedness. The reflection of light is governed by Bragg's law:

$$\lambda_b = \bar{n} P \cos\theta$$

where λ_b is the wavelength of Bragg reflection, $\bar{n}$ is the average refractive index, and P is the length of the helical pitch. The pitch of a chiral nematic is defined as the length traversed by the molecular director $\hat{n}$ on 360° rotation (Figure 4.1a). It is inversely proportional to the concentration $[C]$ as well as the helical twisting power β

83

of the chiral dopant added to the nematic LC to obtain a chiral nematic phase. θ represents the angle of incidence of light. The optical anisotropy of the material gives rise to birefringence Δn which renders the selective reflection band (SRB) with an optical bandwidth $\Delta\lambda$, given by:

$$\Delta\lambda = \Delta n P$$

A change in the length of pitch P translates into a change in the position of the reflection band. When the pitch is in the regime of visible wavelength, such a shift in the wavelength of light reflected is visible to the naked eye. As the color seen of the material is purely structural and it does not involve any electronic excitation, it is photostable (Burgess et al. 2013). These attributes place chiral nematic materials as one of the favorites for the development of low-cost battery-free optical sensors.

Low-cost and easy-to-use optical test strips are attractive in the field of health care (Woltman et al. 2007) for medical diagnostics as they can be used by a common man without specific training and in areas with limited resources. Optical sensors also have applications in real-time (Wang et al. 2015) or time-integrating monitoring of the environment, for example, to detect toxins in water or hazardous chemicals in the surrounding environment. Moreover, as polymer coating labels, they can be of great use for quality control and authentication of consumer products such as food, beverages, drugs, fuel, and cosmetics.

Various optical sensors have been designed by using nonreactive liquid crystalline materials (Mulder et al. 2014). However, this chapter focuses solely on polymer-based optical sensors which have mainly been fabricated by polymerizing functionalized-reactive mesogens to form photonic organic films. Polymerization

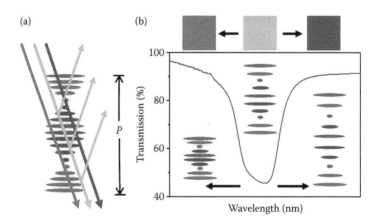

FIGURE 4.1 (a) Schematic of a chiral nematic liquid crystal that reflects circularly polarized light of same handedness as its helicity while transmitting the one with opposite handedness. (b) Swelling of chiral nematic polymer network leads to increase in pitch length thereby redshifting the reflection band, whereas shrinking decreases the pitch length resulting in blueshift of the reflection band.

of reactive mesogens in the chiral nematic phase freezes the helical structure into a polymeric form and provides mechanical strength and thus facilitates fabrication of optical sensors as polymer strips or coatings with ease. Moreover, copolymerization of different reactive mesogens with varying number of end-reactive groups (mono-acrylate and diacrylate) enables tailoring of the desired properties of the material (White and Broer 2015). These photonic films can be constructed to respond to a changing external environment such as temperature, pressure, humidity, and pH or in the presence of a chemical analyte such as alcohol, amino acids, amines, and metal ions. As the number of cholesteric pitches has been fixed due to polymeriza-tion, the optical response arises due to a change in the helical pitch length large enough to cause an alteration in color of the photonic film, perceivable by the naked eye (Figure 4.1b). The response may also be in the form of loss of molecular order leading to disappearance of reflection band.

In this chapter, the aspect of cholesteric liquid crystalline (CLC) polymer films as sensors will be discussed, in which the review by Mulder et al. (2014) has been revised with recent developments in this field. For sensor applications, polymers which consist of hydrogen bonds (H-bonds) have been largely explored for the devel-opment of different kinds of optical sensors in which the observed optical response is due to rupture of the H-bonds. Optical sensors whose working principle is based on absorption and release of water molecules leading to change in pitch will be detailed. Besides, other non-H-bonded optical sensors will also be elaborated in the following section.

4.2 OPTICAL SENSORS

4.2.1 pH Sensors

A pH sensor based on an H-bonded cholesteric polymer composite was first reported by Shibaev et al. (2002). The composite consisted of 1,4-di-(4-(6-acryloxyhexyloxy) benzoyloxy)benzene (DIAB), **1**, as a diacylate cross-linker and 3-methyladipic acid (MAA), **3**, as the chiral dopant. It also contained a polymerizable mesogen, (6-hexa-neoxy-4-benzoic acid) acrylate (HBA), and a nonpolymerizable mesogen, penthylcy-clohexanoic acid (PCA), **2**, with benzoic acid and carboxylic acid functional groups, respectively (Figure 4.2a). On exposing to pH above 7, the reflection band redshifted with as high as 100-nm change in wavelength at pH $= 9$ (Figure 4.2b). The observed effect was attributed to the disruption of the H-bonds due to the neutralization of the acid groups which might have triggered the phase separation of the chiral dop-ant MAA with consequent decrease in its helical twisting power. Another probable reason suggested was the mechanical stress imposed on the polymer matrix due to volumetric changes accompanying the neutralization of the acid groups. Treatment of these films with acidic solutions did not help in restoration of the initial color.

Later, by replacing the nonpolymerizable PCA with a polymerizable mesogen, 4-((6-(acryloyloxy)hexyl)oxy)benzoic acid, **12**, it was possible to partially restore the original color on treatment with an acidic solution (Shibaev et al. 2004). However, several cycles of acid–base treatment of the polymer film led to deterioration of the ester bonds.

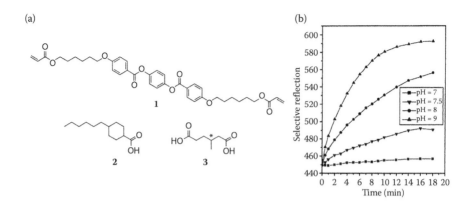

FIGURE 4.2 (a) Chemical structure of the components of the polymer composite. (b) Changes in the position of reflection band on exposing the polymer film to different pH conditions. (Reprinted with permission from Shibaev, P. V. et al. Responsive chiral hydrogen-bonded polymer composites. *Chemistry of Materials* 14 (3): 959–961. Copyright 2002 American Chemical Society.)

In 2015, Stumpel et al. made a novel pH sensor from an interpenetrating network (IPN) of a cholesteric polymer and a poly(acrylic acid) hydrogel (Figure 4.3a) (Stumpel et al. 2015). At pH 9, deprotonation of poly(acrylic acid) occurs and a polymer hydrogel salt, which swells in water, is formed. Absorption of water led to increase in helical pitch resulting in a remarkable redshift of the reflection band by 170 nm (Figure 4.3b). On further treating with a pH 3 buffer, poly(acrylic acid) was formed again and the color reverted to its original position.

4.2.2 AMINO ACID SENSORS

The concept behind the development of a basic amino acid sensor is similar to that of pH sensors. Like the pH sensors, an amino acid sensor also employs an H-bonded CLC polymer and is based on the neutralization of the acid group to form salt. An amino acid sensor was first developed by Shibaev et al. (2006) based on an H-bonded CLC polymer comprising of monomers **1**, **11**, and **12**, and MAA as the chiral dopant. When the cholesteric polymer was exposed to the naturally occurring basic L-arginine solution, neutralization of the benzoic acid took place, which increased the hydrophilicity of the polymer film. As a result, the film swelled, causing a huge redshift of the reflection band by 170 nm (Figure 4.4). Drying of the film did not bring back the reflection band to its original position suggesting the presence of amino acid residues in the polymer matrix. It was also established that the concentration of the monomer **12** with benzoic acid and MAA with carboxylic acid as functional groups plays a role in the response, with higher their concentrations faster is the response.

4.2.3 AMINE SENSORS

With an H-bonded CLC polymer film similar to the one used by Shibaev et al. (2006), an amine sensor can also be made. Stumpel et al. (2014) fabricated a sensor

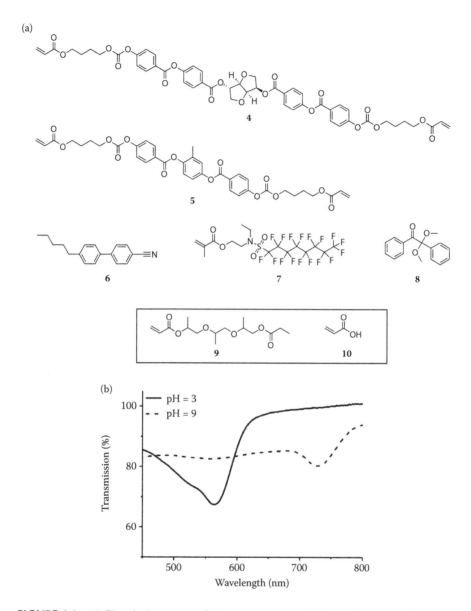

FIGURE 4.3   (a) Chemical structure of the components and (b) UV-Vis transmission spectrum of the IPN of cholesteric polymer and poly(acrylic acid) at pH = 3 and 9. (Stumpel, J. E. et al.: *Advanced Functional Materials*. 3314–3320. 2015. Copyright Wiley-VCH Verlag GmbH & Co. KGaA. Reproduced with permission.)

for trimethylamine (TMA), **15** (Figure 4.5a) which is produced by decaying fish. Interaction of 100% pure TMA with the benzoic acid groups in the polymer film led to the complete loss of cholesteric order and hence disappearance of the reflection band occurred and the polymer film became colorless (Figure 4.5b and c). The response

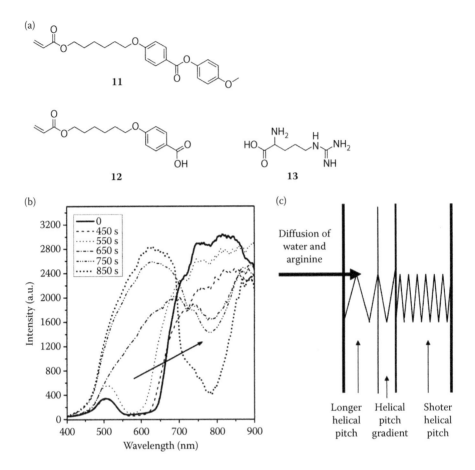

FIGURE 4.4 (a) Chemical structure of the components used. (b) UV-Vis transmission spectra with left-handed circularly polarized light of an H-bonded polymer film when exposed to a solution of L-arginine. (c) Formation of a pitch gradient when arginine solution diffuses through the polymer matrix. (Reprinted with permission from Shibaev, P. V. et al. Color changing cholesteric polymer films sensitive to amino acids. *Macromolecules* 39 (12): 3986–3992. Copyright 2006 American Chemical Society.)

behavior was found to take an S-shape as TMA requires time to diffuse into the polymer film in a non-Fickian manner. On using 13% TMA in water-saturated nitrogen gas, a new redshifted band appeared with reduced intensity (Figure 4.5e). It has been explained that interaction with TMA caused the formation of a carboxylate salt, such as in pH and amino acid sensors (*vide supra*), causing absorption of water by the hygroscopic polymer salt film. Moreover, as a proof of principle, an inkjet-printed polymer film was investigated for its ability to sense the vaporous amine compounds emanated from a decaying fish in a humid environment and indeed, a film exposed to a freshly caught codfish showed a clear color change, similar to TMA in water-saturated nitrogen gas, from green to red after 5 days (Figure 4.5d).

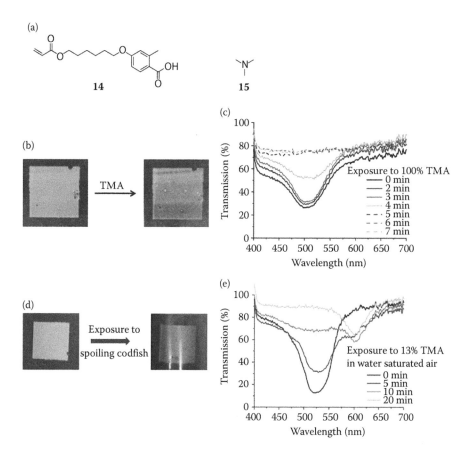

FIGURE 4.5 (a) Structure of monomer 14 used to make amine sensor in combination with the components of an amino acid sensor (P. V. Shibaev et al. 2006). The structure of trimethylamine (TMA) is shown as **15**, (b) film images, and (c) UV-Vis transmission spectra of the cholesteric polymer film on exposure to pure TMA over time. (d) Image of the cholesteric film after 5 days of exposing to a decayed fish. (e) Image of the cholesteric film on exposure to 13% TMA in water-saturated nitrogen gas over time. (Stumpel, J. E. et al.: *Advanced Optical Materials* 459–464. 2014. Copyright Wiley-VCH Verlag GmbH & Co. KGaA. Reproduced with permission.)

4.2.4 Humidity Sensors

So far we have seen that an H-bonded CLC network can be developed into pH, amino acid, or amine optical sensors based on the neutralization of the acid group resulting in change in helical pitch. Herzer et al. (2012) explored the possibility of using hygroscopic carboxylic salt cholesteric polymer film of similar components as was employed by Shibaev et al. (2006) for using as an optical sensor for determining humidity level in the surrounding environment (Herzer et al. 2012). The carboxylic acid salt was obtained by treating a pristine H-bonded CLC film with KOH solution. The hygroscopic film on exposing to a gas-flow chamber with a relative

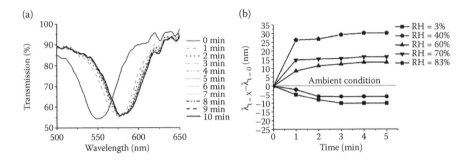

FIGURE 4.6 (a) UV-Vis transmission spectra of the polymer salt film on exposure to 83% RH. (b) Change in the wavelength position of the reflection band of the polymer salt film on exposure to different RH levels. The reflection band position of the film under ambient conditions was considered as the reference. (Reprinted with permission from Herzer, N. et al. Printable optical sensors based on H-bonded supramolecular cholesteric liquid crystal networks. *Journal of the American Chemical Society* 134 (18): 7608–7611. Copyright 2012 American Chemical Society.)

humidity (RH) of 83%, caused the reflection band redshift by 30 nm due to swelling of the polymer film (Figure 4.6a). The polymer salt film responded to a wide range of RH levels and was sensitive to RH as low as 3% (Figure 4.6b). The blueshift of the reflection band observed for lower values of RH is due to lower content of water as compared with the ambient conditions under which the reference reflection band position was taken. The response to humidity is really fast and reversible such that successive exposure of the film to different RH levels gave corresponding optical responses.

The pH sensor which Stumpel et al. had developed from an IPN of cholesteric polymer and hydrogel poly(acrylic acid) can also be designed as a humidity sensor (Stumpel et al. 2015). Deprotonation of poly(acrylic acid) occurred on treatment with KOH solution leading to the formation of a potassium hydrogel salt which is highly hygroscopic. This polymer film was found to respond to different RH levels and an optical response as large as 120 nm was observed when RH changes from RH = 6% to RH = 80% (Figure 4.7).

4.2.5 Alcohol Sensors

Carboxylate salt cholesteric films, besides water, may also attract other polar solvents such as alcohol. Based on an H-bonded CLC mixture consisting of mesogens **5**, **6**, **11**, **12**, and **14** with **4** as the chiral dopant, Chang et al. (2012a,b) fabricated a polymer film that can detect alcohol as well as distinguish between ethanol and methanol. Such sensors are useful to detect, for example, methanol in wine. Cyanobiphenyl mesogen **6** is nonpolymerizable and acts as porogen. Removal of **6** by heating the film caused generation of porosity in the polymer film, which enhances the sensitivity of the film. The polymer film was activated by treating it with an alkaline solution to form the carboxylate salt which then interacts with polar alcohol molecules. The Hildebrand solubility parameter of ethanol (26.5 MPa) is closer to that

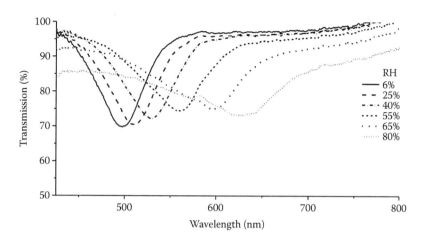

FIGURE 4.7 Optical response of the IPN to different relative humidity (RH) levels. (Stumpel, J. E. et al.: *Advanced Functional Materials* 3314–3320. 2015. Copyright Wiley-VCH Verlag GmbH & Co. KGaA. Reproduced with permission.)

of benzoic acid (21.8 MPa) compared to methanol (29.6 MPa). As a result, exposure of the blue-colored carboxylate salt film to 40% ethanol–60% water led to a large degree of swelling leading to a red color, whereas in the case of exposure to 40% methanol–60% water, the film's color changed to green (Figure 4.8a).

The response of the polymer film was also studied on exposure to varying ratios of ethanol and methanol keeping the total alcohol content at 40% and water at 60% (Figure 4.8b). It was found that as the ratio of ethanol increases, the reflection band shifted more and more toward higher wavelengths. The redshift was more pronounced at lower ratios of ethanol while at higher ratios, the intensity of the reflection is markedly less due to reduced order triggered by the absorption of a large amount of ethanol in the system. The selectivity between ethanol and methanol was found to increase with increase in the number of carboxylic salt sites in the polymer film. The importance of having a porogen in enhancing the sensitivity and selectivity of the film was also established. As can be seen in Figure 4.8c, the polymer film which had contained a porogen gave a more pronounced and different optical response for different ratios of ethanol to methanol for specific alcohol content in water.

4.2.6 METAL ION SENSORS

Crown ethers are well known for their ability to form host–guest complexes. Crown ethers act as host and cations act as guest and depending on the size of the cavity and the ionic radius, they show selectivity toward certain cations that fit well in the cavity. Stroganov et al. (2012) exploited this novel property of crown ethers by developing a metal ion sensor from a cholesteric polymer comprising of mesogens **5** (cross-linker), **16** (nematic LC E48, Merck), **17** (chiral dopant), and **18** (monoacrylate) functionalized with 18-crown-6 moiety (Figure 4.9a). On exposing these films to solutions of Ba^{2+} and K^+ ions, a blueshift of the reflection band was observed.

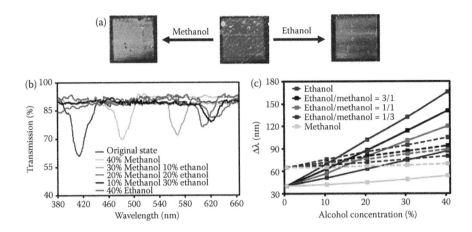

FIGURE 4.8 (a) Images of the carboxylate salt film on exposure to methanol and ethanol. (b) UV-Vis transmission spectra of the carboxylate salt film exposed to varying ratios of ethanol and methanol keeping total alcohol content at 40% and water at 60%. (c) Optical response of the polymer salt film with (solid) and without (dashed) porogen for different concentrations of alcohol with varying ratios of ethanol and methanol in water. (Reprinted with permission from Chang, C.-K. et al. Discrimination of alcohol molecules using hydrogen-bridged cholesteric polymer networks. *Macromolecules* 45 (11): 4550–4555. Copyright 2012b American Chemical Society.)

The response was much higher for Ba^{2+} ions although both K^+ (1.33 Å) and Ba^{2+} (1.34 Å) have similar ionic radii. This result has been attributed to the charge of Ba^{2+} ions which is twice that of K^+ ions. The observed blueshift might have originated due to microphase separation of the crown ether–metal complex which may have triggered the increase in the nonreactive chiral dopant in the LC phase causing further twisting of the helical structure. Another probable reason is the slight collapse of the polymer structure due to the ionic interaction between the positively charged crown ether–metal complex and the counter ions.

Moirangthem et al. (2016) came up with the idea of exploring the metal ion sensing capabilities of the earlier reported H-bonded CLC polymer films. The polymer film was fabricated from a cholesteric mixture which has mesogens with benzoic acid moieties, **12** and **14**, and R-(+)-3-MAA, **3**, as the chiral dopant. Additionally, it also consisted of diacrylate **5** and monoacrylate **11**. MAA, being nonpolymerizable, was easily washed away with an organic solvent and its removal rendered the film more flexible which is prerequisite to observe a high optical response. The H-bonded film, as such, could not bind to the metal ions. It had to be first neutralized with an alkaline solution (KOH) which led to formation of potassium carboxylate salt. As discussed earlier, carboxylate salt is hydrophilic and hence absorbs water causing swelling of the polymer film. What is important to note here is that K^+ ions are labile and therefore, on treatment with various aqueous metal ions ($M^{n+} = Na^+$, Mg^{2+}, Ca^{2+}, Zn^{2+}, Cd^{2+}, and Pb^{2+}), K^+ ions got replaced by the other metal ions. Depending on the hydration capacity of the newly formed M^{n+}–carboxylate complex, excess amount of water was released from the system leading to shrinkage of pitch and

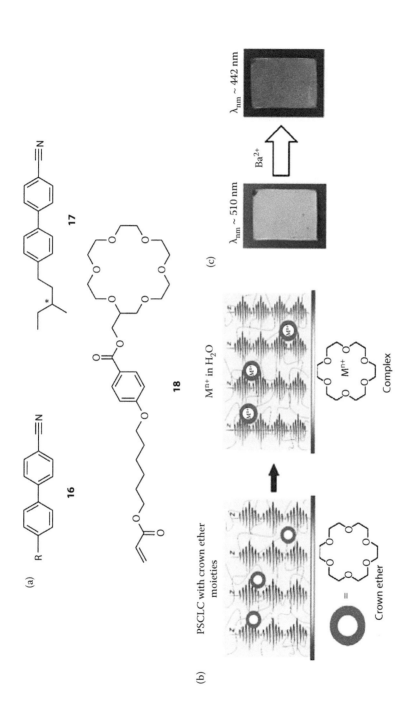

FIGURE 4.9 (a) Structure of the mesogens used to make a metal ion sensor. (b) Schematic illustration of the working principle of a CLC polymer film with crown ether moiety for metal ion sensing. (c) Image of the polymer before and after treatment with Ba²⁺ ions. (Stroganov, V. et al.: *Macromolecular Rapid Communications* 1875–1881. 2012. Copyright Wiley-VCH Verlag GmbH & Co. KGaA. Reproduced with permission.)

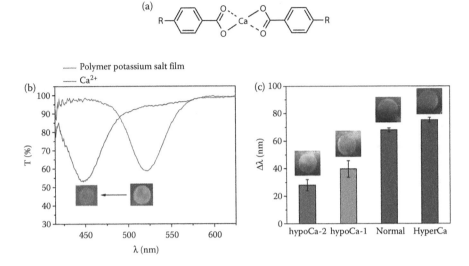

FIGURE 4.10 (a) Schematic illustration of Ca^{2+} binding to planarly aligned benzoate sites forming a planar geometry. Optical response of the potassium carboxylate salt film on exposing (b) to Ca^{2+} ions and (c) to normal serum and serum samples mimicking hypocalcemia and hypercalcemia states. (Moirangthem, M. et al.: *Advanced Functional Materials* 1154–1160. 2016. Copyright Wiley-VCH Verlag GmbH & Co. KGaA. Reproduced with permission.)

hence, blueshift of the reflection band. It was observed that among all the cations, Ca^{2+} gave the maximum optical response ($\Delta\lambda = 70$ nm) (Figure 4.10b).

Selectivity toward Ca^{2+} ions was established by performing two sets of experiment. First, the blueshift of the reflection band observed on exposing a potassium carboxylate salt film to a solution containing a mixture of metal ions including Ca^{2+} was found to be close to the optical response of Ca^{2+} ions alone. Second, treating a Ca^{2+}–carboxylate film with Zn^{2+}, Mg^{2+}, Cd^{2+}, and Na^+ consecutively did not result in any change in the position of the reflection band. These studies showed that Ca^{2+} binds selectively to the polymer film. The observed preference could be explained by the tendency of Ca^{2+} to bind bidentately to benzoate which was facilitated by the planar alignment of the benzoate groups in the polymer film (Figure 4.10a). High optical response of the polymer may be due to the low polarizing ability of Ca^{2+} ions to bind first shell water molecules. The effective concentration range 10^{-4}–10^{-2} M, for which the polymer film is sensitive to Ca^{2+}, makes the film interesting for the detection of calcium in serum. The film, in fact, showed different optical responses for a normal serum and serum mimicking hypocalcemia state as well as hypercalcemia state (Figure 4.10c) acknowledging the potential of using such a polymer salt film as easy-to-use diagnostic test strips.

4.2.7 TEMPERATURE SENSORS

Chen et al. (2011) developed a thermal-responsive cholesteric polymer containing two different chiral dopants, one of which is a nonpolymerizable chiral pyridine derivative, **19**, which can form H-bonds by accepting protons from a mesogen with benzoic acid moieties, **12**, and a polymerizable chiral molecule, **20**. Diacrylate mesogen, **21**, was also employed to act as cross-linker. The HTP of the chiral dopant **19** decreases with increasing temperature. At 30°C, the reflection band of the polymer film was centered at 460 nm (Figure 4.11). However, on heating to 75°C, the HTP reduced significantly due to weakening of the H-bonds and the reflection band redshifted to 560 nm. A marked decrease in the refection intensity was observed as the temperature was close to the clearing point of the cholesteric polymer.

Herzer et al. (2012) studied the possibility of using a humidity-responsive CLC polymer film (described earlier) as a temperature sensor. The reflection band for a water-saturated CLC film was monitored at three different temperatures—room temperature (20°C), in a refrigerator maintained at 4°C, and in a freezer kept at −25°C, and it was found that the reflection band of the film at 20°C was the first to return to its original position in just 10 min (Figure 4.12a); the film kept at 4°C took an hour while only a small blueshift could be seen in case of the film kept at −25°C, which might be attributed to the fact that the reflection band was measured at room temperature which must have led to evaporation of a small amount of water. Inspired by these results, the optical response of the film was monitored within the temperature window of −5°C to +5°C for a time interval of an hour (Figure 4.12b). The films kept at +1°C and +5°C were found to return to the original green color in an hour in agreement with the result obtained earlier and showed time–temperature integrating behavior. Moreover, the CLC mixture could also be inkjet printed on a triacetyl cellulose (TAC) foil and the printed film showed similar optical sensing properties which made it an interesting sensor for recording thermal history.

A temperature sensor can also be developed by using the shape memory behavior of CLC glassy polymers (Davies et al. 2013). The CLC network was made by polymerization of a mixture consisting of a diacrylate, **5**, a monoacrylate, **11**, and monomers with benzoic acid moieties, **12** and **14**, and **4** as the chiral dopant. Mechanical embossing of the polymer film with a spherical metal stamp above its glass transition temperature ($T_g = 50$°C) at 60°C resulted in a spherical indentation of diameter 0.4–0.5 mm and compression of helical pitch that translated into a blueshift of the reflection band by ∼30 nm (Figure 4.13). Reheating the polymer film above its T_g caused the deformed area to recover its original shape accompanied by an irreversible redshift of the reflection band to its initial position. The temperature sensor was found to demonstrate time–temperature integrating behavior between 40°C and 55°C.

A similar approach was used by Benelli et al. (2016) to develop a reversible temperature sensor from a bright green cholesteric polymer containing azobenzene chromophores (Figure 4.14a and c). The polymer exhibited Bragg reflection at 477 nm (Figure 4.14b). Irradiating with UV light below T_g led to increase in the population of *cis*-isomer of the azobenzene causing a decrease in LC order and hence it became transparent (Figure 4.14d). Below the T_g, due to the confinement enforced, the azobenzene did not have enough mobility to isomerize back to the *trans*-form and the configuration stayed locked. It had to be heated above its T_g to regain the initial LC order (Figure 4.14e). The T_g of the polymers designed is notably very high ranging from 90°C to 125°C.

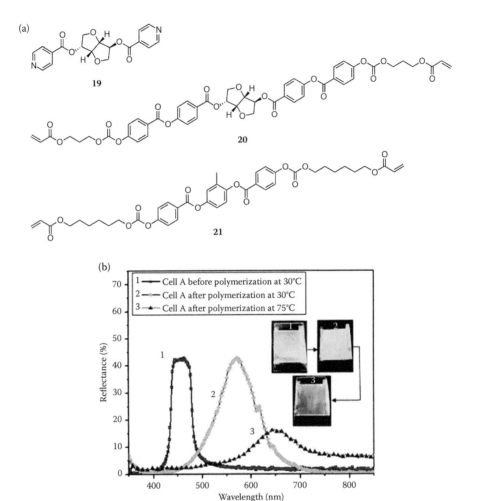

FIGURE 4.11   (a) Structure of the chiral dopants and diacrylate mesogens used to make the thermoresponsive polymer. (b) UV-Vis spectrum of the cholesteric mixture at 30°C (curve 1), and the cholesteric polymer film at 30°C (curve 2) and at 75°C (curve 3). The insets show the photographs of the respective cells. (Chen, Fengjin et al. 2011. Novel photo-polymerizable chiral hydrogen-bonded self-assembled complexes: Preparation, characterization and the utilization as a thermal switching reflective color film. *Journal of Materials Chemistry* 21: 8574–8582. Reproduced by permission of The Royal Society of Chemistry.)

4.2.8   STRAIN SENSORS

A real-time optical strain sensor was developed by Picot et al. (2013) for monitoring uniaxial deformations in oriented polymer films. A cholesteric mixture was first spray coated on a uniaxially aligned polyamide **6** substrate and on photopolymerization, a cross-linked cholesteric polymer was obtained. When a uniaxial extension was applied on the substrate perpendicular to the direction of helical axis of the

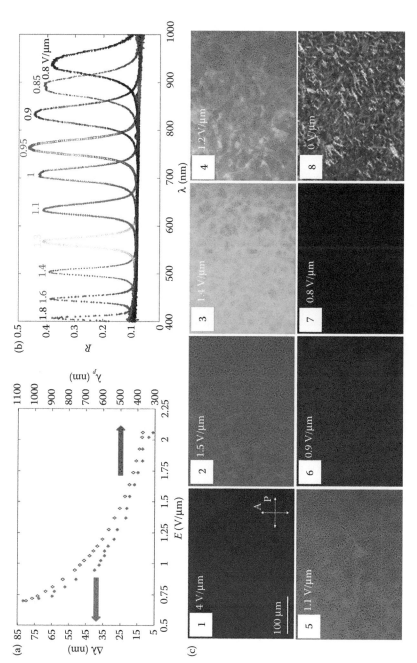

FIGURE 1.8 (a) Wavelength and bandwidth of the selective reflection peak. (b) The reflection spectra. (c) The polarizing optical microscope textures under different amplitudes of the electric field. (Xiang, J. et al.: *Adv. Mater. 3014.* 2015. Copyright Wiley-VCH Verlag GmbH & Co. KGaA. Reproduced with permission.)

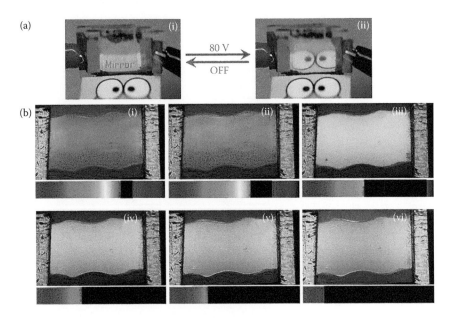

FIGURE 1.11 (a) Photographs of the transmission and reflection transforming mode of the PSCLC at (i) 0 V and (ii) 80 V DC field. (b) The reflection-tunability of the PSCLC is visualized at (i) 0 V, (ii) 15 V, (iii) 30 V, (iv) 60 V, (v) 90 V, and (vi) 110 V. (Reprinted with permission from Lee, K. M. et al. *ACS Photonics* 1, 1033. Copyright 2014. American Chemical Society.)

(a)

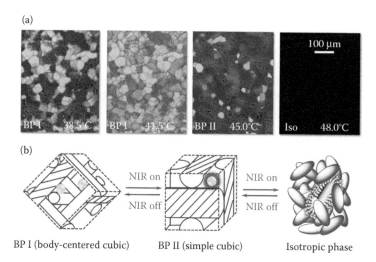

(b)

BP I (body-centered cubic)   BP II (simple cubic)   Isotropic phase

FIGURE 2.14 NIR-light-directed self-organized BP 3D photonic superstructures loaded with gold nanorods. (a) Typical textures of BP I and BP II nanostructures at different temperatures and (b) schematic illustration of the structural transformations under the NIR-light irradiation. (Wang, L. et al. 2015. NIR light-directing self-organized 3D photonic superstructures loaded with anisotropic plasmonic hybrid nanorods. *Chem. Comm.* 51: 15039–15042. Reproduced by permission of The Royal Society of Chemistry.)

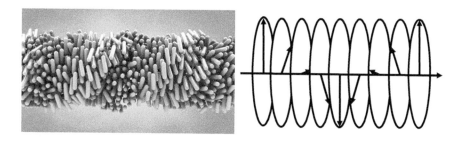

FIGURE 3.1 An illustration of the N* mesophase. On the left is a computer-generated image showing the alignment of individual molecules, while the image on the right shows the helical progression in the alignment of the director. (Image adapted with permission of Yan Liang/BeautifulChemistry.net.)

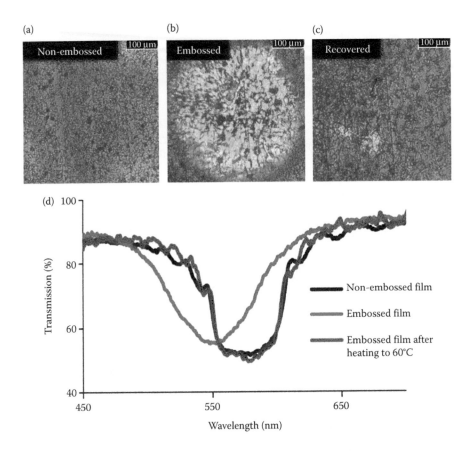

FIGURE 4.13 Microscopy images of (a) nonembossed film, (b) embossed film, and (c) embossed film after heating to 60°C. All the images were taken in the reflection mode. (d) Corresponding UV-Vis transmission spectra. (Davies, D. J. D. et al.: *Advanced Functional Materials* 2723–2727. 2013. Copyright Wiley-VCH Verlag GmbH & Co. KGaA. Reproduced with permission.)

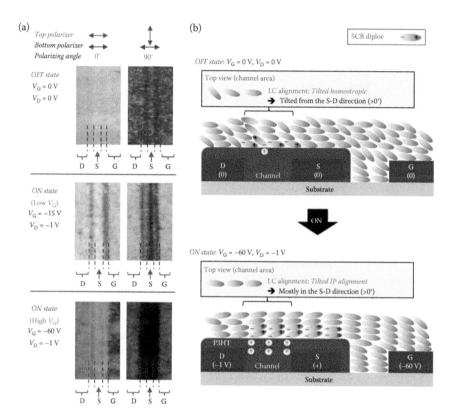

FIGURE 6.8 (a) Optical microscope images for the LC-*g*-OFET devices according to the drain and gate voltages: (left) linear polarization state (0°), (right) cross-polarization state (90°). (b) Illustration for the movement of LC molecules according to the drain and gate voltages: (top) OFF state, (bottom) ON state. (A~B: Reprinted with permission from Seo, J. et al. 2015. Liquid crystal-gatedorganic field-effect transistors with in-plane drain–source–gate electrode structure. *ACS Appl. Mater. Interfaces* 7: 504–510. Copyright 2015 American Chemical Society.)

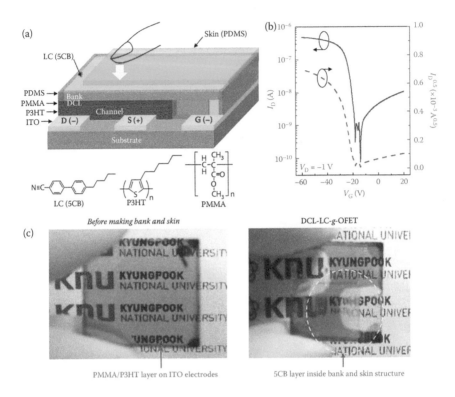

(a)

Skin (PDMS)

LC (5CB)

PDMS →
PMMA →
P3HT →
ITO →

Bank
DCL
Channel

D (−)   S (+)   G (−)

Substrate

N≡C―⟨ ⟩―⟨ ⟩―〜

LC (5CB)

P3HT

PMMA

(b)

$V_D = -1\ V$

I_D (A)

V_G (V)

$I_D^{0.5}\ (\times 10^{-3}\ A^{0.5})$

(c)

Before making bank and skin

DCL-LC-*g*-OFET

PMMA/P3HT layer on ITO electrodes

5CB layer inside bank and skin structure

FIGURE 6.15 (a) Illustration for the device structure of DCL-LC-*g*-OFET device with the poly(dimethyl siloxane) (PDMS) protection skin layer: The width and length of ITO electrodes (D, S, and G) were 18 μm and 3 mm, respectively. (b) Transfer characteristics of the DCL-LC-*g*-OFET device at $V_D = -1$ V. (c) Photographs for the DCL-LC-*g*-OFET device with (left) and without (right) the PDMS bank and skin parts. (A~C: Reprinted from *Org. Electron.*, 28, Seo, J. et al., Physical force-sensitive touch responses in liquid crystal-gated-organic field-effect transistors with polymer dipole control layers., 184–188, Copyright 2016, with permission from Elsevier.)

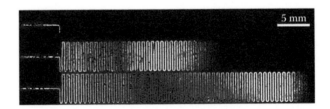

5 mm

FIGURE 7.4 Correlations between the length of the bright LC regions and the concentration of anti-IgG. From the top, the concentration of anti-IgG was 0.02, 0.05, and 0.08 mg/mL, respectively.

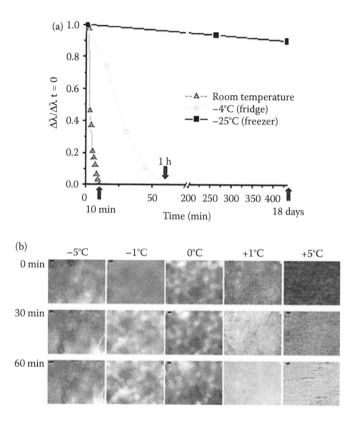

FIGURE 4.12 (a) Change in the wavelength of the reflection of the water-saturated films kept at room temperature, +4°C and −25°C with reference to the ambient condition as a function of time. (b) Microscopy images of the films at −5°C, −1°C, 0°C, +1°C, +5°C captured in reflection mode without cross-polarizers after 0, 30, and 60 min. (Reprinted with permission from Herzer, N. et al. Printable optical sensors based on H-bonded supramolecular cholesteric liquid crystal networks. *Journal of the American Chemical Society* 134 (18): 7608–7611. Copyright 2012 American Chemical Society.)

coated cholesteric polymer, the cholesteric polymer expanded in the xy plane with consequent shrinkage of pitch. A strain of 13% resulted in blueshift of the reflection band from ~595 nm to ~555 nm (Figure 4.15). On average, the sensitivity of the bilayer was approximately 3 nm/% strain. The close agreement of the mechanical response of the polymer substrate to strain and the optical response of the cholesteric polymer coating illustrates the potential of the developed sensor for real-time strain sensing.

4.3 CONCLUSION

Several optical sensors for diverse applications have been developed based on the CLC polymer films. The optical response of the polymer films is based on the ability of the cholesteric phase to reflect light, which renders such sensors battery-free.

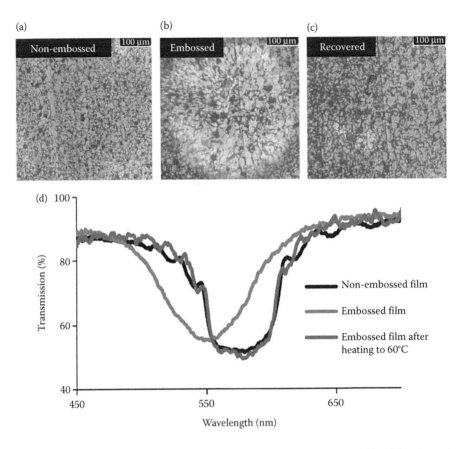

FIGURE 4.13 (**See color insert.**) Microscopy images of (a) nonembossed film, (b) embossed film, and (c) embossed film after heating to 60°C. All the images were taken in the reflection mode. (d) Corresponding UV-Vis transmission spectra. (Davies, D. J. D. et al.: *Advanced Functional Materials* 2723–2727. 2013. Copyright Wiley-VCH Verlag GmbH & Co. KGaA. Reproduced with permission.)

For the optical response to be clearly visible to the naked eye, it is necessary that the optical sensor shows large response and this has been achieved by utilizing different combinations of reactive and nonreactive mesogens thereby tuning the flexibility of the polymer network, as desired. These sensors are low cost and have a response discernable without any aid making them user-friendly which is crucial for these technologies to reach out to common people.

It is, however, important to mention that chemical sensors reported so far suffer from the drawback of being cross-selective and cross-sensitive; for example, an amino acid sensor is also sensitive to amines or any other chemical analyte which can trigger the rupture of H-bonds responsible for the optical response. Moreover, sensors for physical parameters such as temperature are not yet in the regime of actual relevance. A temperature sensor which can be useful as smart labels in packaging

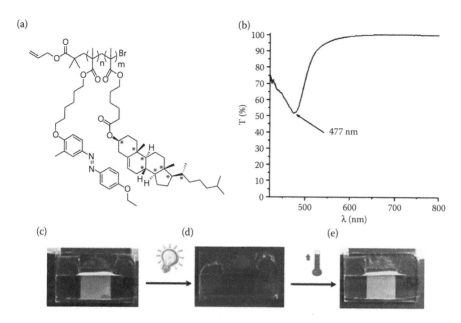

FIGURE 4.14   (a) Molecular structure (n:m = 52:48) and (c) film image of the cholesteric polymer containing azobenzene chromophores. (b) UV-Vis spectra of the cholesteric polymer without azoaromatic counits contribution. (d) UV irradiation leads to *trans–cis* isomerization. (e) The *cis*-isomer of azobenzene relaxes to the thermodynamically more stable *trans*-form on heating above the glass transition temperature. (Reprinted from *Dyes and Pigments*, 126, Benelli, T. et al., Supramolecular ordered photochromic cholesteric polymers as smart labels for thermal monitoring applications, 8–19, Copyright 2016, with permission from Elsevier.)

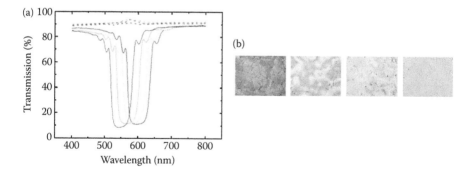

FIGURE 4.15   (a) UV-Vis transmission spectra and (b) the corresponding microscopy images of the bilayer film shows change in color on increasing the applied strain from 0% to 13%. (Picot, O. T. et al. 2013. A real time optical strain sensor based on a cholesteric liquid crystal network. *RSC Advances* 3 (41): 18794–18798. Reproduced by permission of The Royal Society of Chemistry.)

industries needs an operating window of 0–8°C, whereas the sensors developed so far work above room temperature. Although these optical sensors are still far from real usage in daily life, cholesteric polymers definitely are promising materials and much is left to be explored in this direction.

REFERENCES

Benelli, T., L. Mazzocchetti, G. Mazzotti, F. Paris, E. Salatelli, and L. Giorgini. 2016. Supramolecular ordered photochromic cholesteric polymers as smart labels for thermal monitoring applications. *Dyes and Pigments* 126: 8–19, doi: 10.1016/j.dyepig.2015.11.009.

Burgess, I. B., M. Lončar, and J. Aizenberg. 2013. Structural colour in colourimetric sensors and indicators. *Journal of Materials Chemistry C* 1 (38): 6075–6086, doi: 10.1039/c3tc30919c.

Chang, C.-K., C. M. W. Bastiaansen, D. J. Broer, and Hui Lung Kuo. 2012a. Alcohol-responsive, hydrogen-bonded, cholesteric liquid-crystal networks. *Advanced Functional Materials* 22 (13): 2855–2859, doi: 10.1002/adfm.201200362.

Chang, C.-K., C. W. M. Bastiaansen, D. J. Broer, and Hui Lung Kuo. 2012b. Discrimination of alcohol molecules using hydrogen-bridged cholesteric polymer networks. *Macromolecules* 45 (11): 4550–4555, doi: 10.1021/ma3007152.

Chen, Fengjin, Jinbao Guo, Zhijian Qu, and Jie Wei. 2011. Novel photo-polymerizable chiral hydrogen-bonded self-assembled complexes: Preparation, characterization and the utilization as a thermal switching reflective color film. *Journal of Materials Chemistry* 21: 8574–8582, doi: 10.1039/c0jm03810e.

Davies, D. J. D., A. R. Vaccaro, S. M. Morris, N. Herzer, A. P. H. J. Schenning, and C. W. M. Bastiaansen. 2013. A printable optical time-temperature integrator based on shape memory in a chiral nematic polymer network. *Advanced Functional Materials* 23 (21): 2723–2727, doi: 10.1002/adfm.201202774.

Fenzl, C., T. Hirsch, and O. S. Wolfbeis. 2014. Photonic crystals for chemical sensing and biosensing. *Angewandte Chemie International Edition* 53 (13): 3318–3335, doi: 10.1002/anie.201307828.

Ge, Jianping, and Yadong Yin. 2011. Responsive photonic crystals. *Angewandte Chemie International Edition* 50 (7): 1492–1522, doi: 10.1002/anie.201307828.

Herzer, N., H. Guneysu, D. J. D. Davies, D. Yildirim, A. R. Vaccaro, D. J. Broer, C. W. M. Bastiaansen, and A. P. H. J. Schenning. 2012. Printable optical sensors based on H-bonded supramolecular cholesteric liquid crystal networks. *Journal of the American Chemical Society* 134 (18): 7608–7611, doi: 10.1021/ja301845n.

Moirangthem, M., R. Arts, M. Merkx, and A. P. H. J. Schenning. 2016. An optical sensor based on a photonic polymer film to detect calcium in serum. *Advanced Functional Materials* 26 (8): 1154–1160, doi: 10.1002/adfm.201504534.

Mulder, D.-J., A. P. H. J. Schenning, and C. Bastiaansen. 2014. Chiral-nematic liquid crystals as one dimensional photonic materials in optical sensors. *Journal of Materials Chemistry C* 2 (33): 6695–6705, doi: 10.1039/c4tc00785a.

Picot, O. T., M. Dai, E. Billoti, D. J. Broer, T. Peijs, and C. W. M. Bastiaansen. 2013. A real time optical strain sensor based on a cholesteric liquid crystal network. *RSC Advances* 3 (41): 18794–18798, doi: 10.1039/c3ra42986e.

Shibaev, P. V., D. Chiappetta, R. Lea Sanford, P. Palffy-Muhoray, M. Moreira, W. Cao, and M. M. Green. 2006. Color changing cholesteric polymer films sensitive to amino acids. *Macromolecules* 39 (12): 3986–3992, doi: 10.1021/ma052046o.

Shibaev, P. V., J. Madsen, and A. Z. Genack. 2004. Lasing and narrowing of spontaneous emission from responsive cholesteric films. *Chemistry of Materials* 16 (8): 1397–1399, doi: 10.1021/cm0305812.

Shibaev, P. V., K. Schaumburg, and V. Plaksin. 2002. Responsive chiral hydrogen-bonded polymer composites. *Chemistry of Materials* 14 (3): 959–961, doi: 10.1021/cm011510a.

Stroganov, V., A. Ryabchun, A. Bobrovsky, and V. Shibaev. 2012. A novel type of crown ether-containing metal ions optical sensors based on polymer-stabilized cholesteric liquid crystalline films. *Macromolecular Rapid Communications* 33 (21): 1875–1881, doi: 10.1002/marc.201200392.

Stumpel, J. E., D. J. Broer, and A. P. H. J. Schenning. 2014. Stimuli-responsive photonic polymer coatings. *Chemical Communications (Cambridge, England)* 50 (100). *Royal Society of Chemistry*: 15839–15848, doi: 10.1039/c4cc05072j.

Stumpel, J. E., E. R. Gil, A. B. Spoelstra, C. W. M. Bastiaansen, D. J. Broer, and A. P. H. J. Schenning. 2015. Stimuli-responsive materials based on interpenetrating polymer liquid crystal hydrogels. *Advanced Functional Materials* 25 (22): 3314–3320, doi: 10.1002/adfm.201500745.

Stumpel, J. E., C. Wouters, N. Herzer, J. Ziegler, D. J. Broer, C. W. M. Bastiaansen, and A. P. H. J. Schenning. 2014. An optical sensor for volatile amines based on an inkjet-printed, hydrogen-bonded, cholesteric liquid crystalline film. *Advanced Optical Materials* 2 (5): 459–464, doi: 10.1002/adom.201300516.

Wang, Ding, Soo-Young Park, and Inn-Kyu Kang. 2015. Liquid crystals: Emerging materials for use in real-time detection applications. *Journal of Materials Chemistry C* 3 (35). Royal Society of Chemistry: 9038–9047, doi: 10.1039/C5TC01321F.

White, T. J., and D. J. Broer. 2015. Programmable and adaptive mechanics with liquid crystal polymer networks and elastomers. *Nature Materials* 14 (11): 1087–1098, doi:10.1038/nmat4433.

White, T. J., M. E. McConney, and T. J. Bunning. 2010. Dynamic color in stimuli-responsive cholesteric liquid crystals. *Journal of Materials Chemistry* 20 (44): 9832, doi: 10.1039/c0jm00843e.

Woltman, S. J., G. D. Jay, and G. P. Crawford. 2007. Liquid-crystal materials find a new order in biomedical applications. *Nature Materials* 6 (12): 929–938, doi: 10.1038/nmat2010.

5 All-Electrical Liquid Crystal Sensors

Juan Carlos Torres Zafra, Braulio García-Cámara,
Carlos Marcos, Isabel Pérez Garcilópez,
Virginia Urruchi, and José M. Sánchez-Pena

CONTENTS

5.1 Introduction 103
5.2 Experimental Devices... 105
    5.2.1 Fabrication of LC Cells... 105
    5.2.2 Experimental Permittivity of LC Devices... 107
    5.2.3 Electrical Equivalent Circuit 108
    5.2.4 Temperature Dependence of the Electrical Equivalent Circuit........ 109
5.3 LC-Based Square Waveform Generator as Temperature Sensor
    (Temperature–Frequency Converter) 111
    5.3.1 Temperature–Frequency Converter Design... 111
    5.3.2 Experimental Characterization of the LC Cell................................... 112
    5.3.3 Temperature–Frequency Converter Characterization 113
5.4 LC-Based Phase Detector as Temperature Sensor (Temperature-Phase
    Converter) 116
5.5 Conclusions... 120
References.. 120

5.1 INTRODUCTION

One of the main tasks of current technology is the sensing. We should continuously check the value of a myriad of parameters in a large number of complex systems, ranging from the strain of the components of an aircraft (Gil-Garcia et al. 2015) to the DNA sequence variations (Hahm and Lieber 2004). Thus, sensors are a key point in current world, in particular since the introduction of the concepts of Internet-of-things (IoT) (Borges et al. 2015) and smart cities (Perera et al. 2013). For each interesting parameter, we can find several kinds of sensors, depending on the physical properties or phenomena that are used in their design. In this chapter, we focus on sensors using the electronic properties of a very interesting material: liquid crystals (LCs).

LCs are composed of nanometric organic elongated molecules. This rod-like molecules exhibit orientational order in a particular direction (De Gennes 1975) and they have an intrinsic anisotropy on their main properties (Blinov and Chigrinov 1994):

mechanical, electrical, magnetic, or optical. In particular, their electrical and optical anisotropies, which are easily modified by applying an external stimulus, such as an electric field, make them unique for display applications (Yang and Wu 2006).

LCs are well known since decades, and their properties are well described and characterized in a large number of works. This shows that LCs are used in a wide range of applications. Specifically, LC displays (LCDs) are their best known application. Currently, we can find an LCD in several devices such as mobile phones or digital camera. Furthermore, other applications based on the optical anisotropy of LCs have emerged and the knowledge and technological capabilities to manipulate them has increased. Some of the remarkable ones are, for instance, wavelength-tunable filters (Beeckman et al. 2009), variable optical attenuators (De Gennes 1975), spatial light modulators (Crossland et al. 2000), optical multiplexers (Lallana et al. 2006), or optical adaptative systems (Yao et al. 2014). In addition, LCs are now being used in new research studies in photonics such as Plasmonics (Khatua et al. 2011; Si et al. 2014) and metamaterials (Zhao et al. 2007).

It can be concluded from the previous list that the majority of LC-based devices use the beneficial optical properties of LCs to operate. On the other hand, the electric anisotropy of LC is hardly exploited, although it can be quite interesting for use in certain devices. For instance, in previous works, we showed the use of LC cells in electronic devices such as a sinusoidal oscillator (Pérez et al. 2007), a phase-locked loop (PLL) (Marcos et al. 2011), or a tunable series–parallel resonator (Torres et al. 2012).

On this basis, this chapter is devoted to study the use of the electrical properties of LC in the designing of new devices. In particular, and focusing on the field of sensing, LCs are present in several temperature sensors. Temperature is one of the most important parameters in diverse fields ranging from industry to biological processes. In addition, they require an accurate measurement due to the strong influence of temperature on their operations. In this sense, LCs present an important sensitivity of their main optical properties to any temperature deviation. This is represented by a high thermo-optic coefficient (Li et al. 2005). Furthermore, the electric properties of LC are also dependent on temperature; thus, they can be also used for designing new sensors. Consequently, this work will show two designs of an *all-electric* temperature sensor. Both proposals take the advantage of the electrical properties of LC, in particular the temperature dependence of the impedance of an LC cell, to obtain accurate estimations of temperature.

This chapter is organized as follows: Section 5.2 is devoted to the fabrication and the electric characterization of the experimental LC cells. In particular, we propose a simple electrical equivalent circuit, composed only of resistors and capacitors. From a practical point of view, this involves an intuitive way of obtaining the electric response of the LC cells. Based on the temperature dependence of the nominal values of these equivalent components, we demonstrate two complex electronic devices that act as sensitive temperature dependent sensors. Section 5.3 shows a simple square waveform generator whose oscillating frequency is very sensitive to temperature; the second system for sensing is implemented by a phase detector (Section 5.4) that can also act as a temperature sensor. Finally, Section 5.5 summarizes the designed sensors, highlighting their main advantages with respect to current devices.

5.2 EXPERIMENTAL DEVICES

This section is devoted to explain the fabrication procedure of the experimental devices we followed as well as their electric characterization.

5.2.1 FABRICATION OF LC CELLS

A nematic liquid crystal (NLC) cell consists of a thin layer of the material sandwiched between two parallel transparent plates with a conductive layer in their inner surfaces. In this procedure, a layer of indium-tin-oxide (ITO) is used as a conductive medium; however, other compounds found in the literature can also be used. One of the most critical manufacturing parameters of LC cells is the thickness. Cell gap is accurately fixed using spacers distributed along the plate area. In order to analyze the influence of this variable on the sensing properties of the device, cells with different thicknesses were prepared. The optimal value of this parameter for temperature sensing has been verified.

On the other hand, alignment films deposited over ITO electrodes are needed to ensure the molecular orientation of LC on the substrates. Basically, LC cells can be manufactured using two alignment processes: planar and homeotropic. Alignment is also a key parameter of these devices. For a planar alignment, LC molecules are oriented near parallel to the surface of the substrates (Figure 5.1a), using rubbed polyimides for the alignment layer. For LC materials with positive dielectric anisotropy, the presence of an external electric field induces a reorientation of the molecules so that their long axes tend to align parallel to this field (Figure 5.1b).

Furthermore, with a homeotropic alignment, LC molecules are placed perpendicular to the surface of the substrates (Figure 5.2a) due to the presence of a monolayer of surfactants (silane). In this case, the orientation of the molecules cannot be altered by the application of an external electric field (Figure 5.2b), for LCs with positive dielectric anisotropy ($\Delta\varepsilon > 0$).

It is well known that NLCs are dielectric materials whose dielectric permittivity (hereafter permittivity) depends on the orientation of molecules due to the anisotropy of their electric properties. The extreme values of electric permittivity of an LC are $\varepsilon_\perp$ and $\varepsilon_\parallel$ (Figure 5.2c). For a planar alignment, when no electric field is applied to the LC cell, the molecules will remain parallel to the substrates. Then, the effective permittivity measured will be near $\varepsilon_\perp$. However, if a strong electric field is applied,

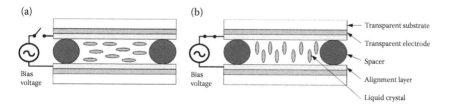

FIGURE 5.1 Schematic illustration of planar alignment in nematic liquid crystal cells (a) and molecular reorientation with an external electric field applied (b).

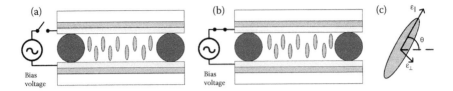

FIGURE 5.2 (a) and (b) correspond with the schematic illustration of homeotropic align-ment in nematic liquid crystal cells without or with an external electric field applied, respec-tively. (c) Scheme of the extreme values of the effective electric permittivity of a liquid crystal.

the molecules rotate to the vertical position (the molecular director tends to point parallel to the electric field direction); therefore, in this case, the measured permit-tivity will be nearly $\varepsilon_{\parallel}$. Any electric field with intermediate values can provide a permittivity between $\varepsilon_{\perp}$ and $\varepsilon_{\parallel}$ (De Gennes 1975).

On the other hand, when a homeotropic alignment is considered, the molecular orientation cannot be altered for any external electric field, as commented above; thus, the effective permittivity remains nearly $\varepsilon_{\parallel}$. If the material has a positive anisot-ropy ($\Delta\varepsilon > 0$) as we are considering, the parallel effective permittivity is greater than the perpendicular one ($\varepsilon_{\parallel} > \varepsilon_{\perp}$) (Kelly and O'Neill 2001).

In addition to the alignment process's dependence, permittivity also depends on the temperature. This is shown in Figure 5.3 for an LC material with positive dielec-tric anisotropy. As can be seen in this figure, $\varepsilon_{\perp}$ and $\varepsilon_{\parallel}$ depend on temperature, which is below the clearing temperature (T_C). Above T_C, the material reaches its isotropic phase and the dielectric anisotropy disappears ($\varepsilon_{\perp} = \varepsilon_{\parallel} = \varepsilon_{iso}$). Besides, when the material is in the nematic mesophase, temperature dependence is not equal for both LC permittivities. The parallel permittivity has a greater variation with temperature than the perpendicular one. The research presented in this work is based on the influ-ence of temperature on a NLC with positive dielectric anisotropy.

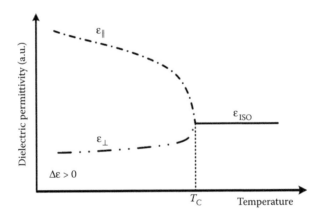

FIGURE 5.3 Typical temperature dependence of the permittivity for a nematic liquid crys-tal with positive anisotropy.

5.2.2 Experimental Permittivity of LC Devices

Due to the interest on the electrical properties of the devices, the first checked parameter of our experimental cells was the effective permittivity, both the real (ε') and the imaginary (ε'') parts. In addition, both kinds of alignments, planar and homeotropic, at room temperature ($\sim$30°C) were considered (Figure 5.4). One of the most common ways to obtain this parameter is through impedance measurements. The impedance (magnitude and phase) of our LC cells was measured with an impedance analyzer (SOLARTRON 1260) using a sinusoidal voltage signal with 100 mVrms (below the threshold voltage that induces the molecular reorientation) and a frequency sweep ranging from 100 Hz to 10 MHz. Besides, in order to analyze properly the influence of temperature on the measured parameters, cells were placed in a programmable environmental chamber (DYCOMETAL CCK-40/180) to ensure a stable temperature during the measurement process.

Although a large number of cells were manufactured and characterized, we only show the results of some of them to ensure clarity. In this case and in order to analyze the frequency dependence of $\varepsilon_\perp$ and $\varepsilon_\parallel$, we have considered an LC cell with each alignment method. Cell 5 has a homeotropic alignment that permits to evaluate the behavior of $\varepsilon_\parallel$ and the molecules of cell 6 are oriented parallel to the substrate (planar alignment) to analyze $\varepsilon_\perp$. In both cases, the cell thickness is 6.3 μm. Results of the real and imaginary parts of the permittivity for planar and homeotropic alignment processes at room temperature ($\sim$30°C) are shown in Figure 5.4.

From Figure 5.4, a similar behavior is observed for both planar and homeotropic cells. However, due to the positive dielectric anisotropy, the values of parallel

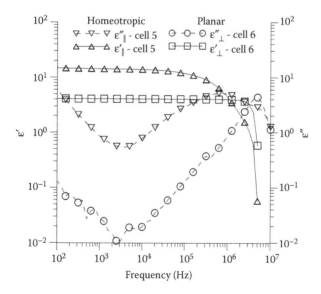

FIGURE 5.4 Experimental data of the complex permittivity of NLC cells as a function of the frequency of an applied electric field and for both planar and homeotropic cells at a room temperature (30°C).

permittivity are greater than perpendicular permittivity, both for the real and imaginary parts. Low frequencies ($f < 100$ Hz) are discarded for these measurements because low electric fields can induce degradation of the LC material due to the absorption of ion charges and generation of electric fields on the electrode layers (Perlmutter et al. 1996; Thurston et al. 1984).

At high frequencies, the imaginary part of permittivity (ε'') presents a maximum. This corresponds to an absorption peak due to the dipolar relaxation. While it appears at 8 MHz for the planar cell (cell 6), for the homeotropic one (cell 5) the maximum is located at $\sim$800 kHz. A feasible explanation of this fact is that less energy is required to produce molecular motion in planar alignment than in homeotropic alignment (Costa et al. 2006). On the other hand, within the medium frequency range, the curves present two regions, with a different dielectric behavior, separated by the minimum reached by the imaginary part of permittivity at about 3 kHz, for both samples (Raistrick et al. 2005). These minima will be analyzed from an impedance point of view below.

5.2.3 ELECTRICAL EQUIVALENT CIRCUIT

Based on the previous results, LC cells are electrically equivalent to a circuit like that plotted in Figure 5.5 in a wide frequency range that depends on manufacturing parameters of the device (100 Hz $< f < 10^5$ Hz in the samples under study). This model includes a capacitor C representing dipolar polarization and a resistor R modeling the mobility of free charges and the dipolar displacement inside the device. In addition, the influence of electrodes is represented with an additional resistor, Rs. However, the effect of this component is only appreciable at high frequencies ($f > 3$ MHz, in the present case).

The value of each component in this circuit as a function of the frequency could be inferred from the complex permittivity of Figure 5.4 by means of Equations 5.1 and 5.2.

$$R(\omega) = \frac{1}{\omega \cdot \varepsilon'' \cdot C_0} \tag{5.1}$$

$$C = \varepsilon' \cdot C_0. \tag{5.2}$$

C_0 being the vacuum device's capacitance.

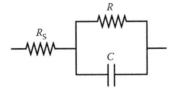

FIGURE 5.5   Electrical equivalent circuit of an LC cell in a wide frequency range. Equivalent electric circuit of an LC cell for a frequency range of 100–10^5 Hz.

5.2.4 Temperature Dependence of the Electrical Equivalent Circuit

The permittivity of NLCs strongly depends on the temperature, among other parameters. In NLCs with positive dielectric anisotropy this dependence is greater for the parallel component ($\varepsilon_{\parallel}$) than for the perpendicular one ($\varepsilon_{\perp}$), as observed in Figure 5.3 (Thurston et al. 1984). This dependence could be modeled through variation of the components of the electrical equivalent circuit. The temperature dependence is noticed in both the resistor, R, and the capacitor, C, due to the variation of both the mobility and the density of the ions as the temperature changes. In order to obtain a simplified temperature dependence, only an ideal capacitive behavior of the LC device has been considered. The importance of this assumption lies on the fact that a dominant capacitance behavior involves minimum power consumption. This is a key point for future devices. In addition, negligible conductivity is also obtained under this assumption. However, this is not suitable for any frequency. The frequency range at which the system acts as an ideal capacitor is that in which the phase impedance of the NLC cell remains close to $-90°$.

Figure 5.6 shows both the modulus and the phase of the impedance of a homeotropic cell as a function of frequency and at several different temperatures. As can be seen, this range is not very large, and it also depends on the temperature.

This temperature dependence of the equivalent capacitance is shown in Figure 5.7. The values of the capacitance inferred from the impedance analysis of cells 5 and 6 are plotted as a function of the temperature, from 0°C to 80°C, and considering a

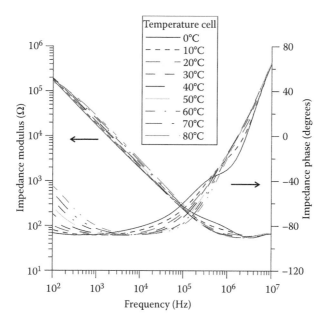

FIGURE 5.6 Experimental complex impedance of a homeotropic NLC cell as function of frequency for a temperature range of 0–80°C.

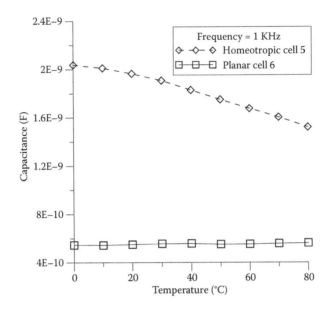

FIGURE 5.7 Variation of capacitance (C) as a function of the external frequency for different temperature values for NLC cell 5 (homeotropic cell) and NLC cell 6 (planar cell).

frequency of 1 kHz, where the impedance analysis reveals an ideal capacitive behavior (phase impedance $\approx$−90°) for all the temperatures checked. As was expected, the capacitance in both cases decreases with temperature because the mobility and the numerical density of the ions increase with temperature. However, while in the planar cell this variation is ca. −0.25 pF/°C, in the homeotropic one, it is around 35 times higher ($\sim$ −8.75 pF/°C). Thus, we can conclude that in the absence of an applied external voltage, devices with a homeotropic alignment are around 35 times more sensible to temperature change.

In order to make a detailed analysis of this behavior, the dependence of the equivalent capacitance on temperature could be expressed as:

$$S_C = \frac{\varepsilon_0 \cdot S}{d} \cdot \frac{\partial \varepsilon'}{\partial T}. \tag{5.3}$$

S_C is the slope of this dependence in pF/°C. ε_0 is the vacuum permittivity, S is the effective area of the electrodes, and d is the cell thickness.

As it can be seen, this dependence can be also controlled through several parameters. For instance, by decreasing the thickness of the device, the sensitivity could be increased linearly. Figure 5.8 shows the sensitivity S_C of two homeotropic cells with different thicknesses, 1.5 μm (cell 4) and 6.3 μm (cell 5), and considering a frequency of 1 kHz. These experimental data evidence this relation, showing that the variation of capacitance, when the temperature changes, is larger in cell 4 ($d = 1.5$ μm) than in cell 5 ($d = 6.3$ μm).

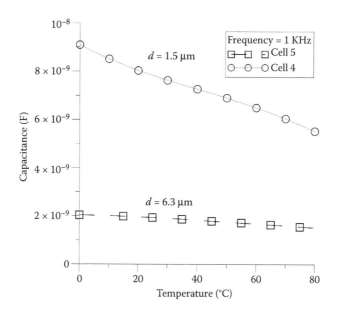

FIGURE 5.8 Temperature dependence of the liquid crystal capacitance taken in homeotropic cells with different thickness: cell 5 (thickness = 6.3 μm) and cell 4 (thickness = 1.5 μm).

5.3 LC-BASED SQUARE WAVEFORM GENERATOR AS TEMPERATURE SENSOR (TEMPERATURE– FREQUENCY CONVERTER)

Previous results demonstrate the possibility to design a temperature sensor based on the temperature dependence of an NLC cell capacitance. A new temperature-sensing electric system based on this dependence is presented.

5.3.1 Temperature–Frequency Converter Design

The proposed system consists of a square wave generator circuit where a planar NLC cell is used as the sensing element via the temperature dependence capacitance. This capacitance controls the output frequency of the waveform generator. Hence, a temperature variation will change the frequency, particularly of a square wave.

The electronic circuit is based on an astable multivibrator circuit shown in Figure 5.9a. The astable multivibrator circuit is able to generate a square output signal, whose amplitude switches between two symmetrical levels (the high level with a voltage of $V+$ and the low level of $V-$). The output square wave is a 50% duty cycle signal; therefore, the output signal maintains a high level during 50% of the signal period, and for the remaining time the output level is low.

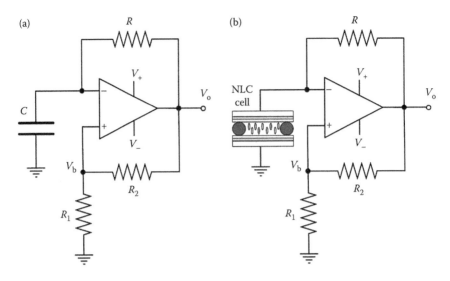

FIGURE 5.9 (a) Multivibrator oscillator circuit. (b) Multivibrator circuit with NLC cell.

The astable multivibrator circuit is implemented with one operational amplifier (op-amp), one capacitor (C), and three resistors (R, R_1, and R_2). The time period of the output signal, T, is given by the following equation (Rashid 1998):

$$T = 2 \cdot R \cdot C \cdot \ln\left(\frac{2R_1 + R_2}{R_2}\right). \tag{5.4}$$

Therefore, time period, T, depends only on the value of the passive components (resistors and capacitor) of the circuit. The temperature–frequency converter is obtained by a simple modification of the astable multivibrator circuit. When the capacitor is replaced by an NLC cell (Figure 5.9b), variation in the output signal frequency depends only on the NLC cell temperature.

5.3.2 EXPERIMENTAL CHARACTERIZATION OF THE LC CELL

Cell 6 with a planar alignment was used in these experiments. Impedance measurements of this cell were done using the impedance analyzer. A sinusoidal voltage below the threshold voltage was applied in a temperature range from 0°C to 80°C. As it can be seen in Figure 5.10, the frequency at which the NLC cell can be assumed to be a pure capacitance, for the temperature range checked, is around 5 kHz.

As it was mentioned earlier, the electrical behavior of the experimental cells was checked by measuring permittivity as a function of frequency, applying a low external voltage, below the threshold voltage that induces molecular reorientation in the LC material. However, in planar cells, a high electric field can provide an effective permittivity between $\varepsilon_\perp$ and $\varepsilon_\parallel$. In this case, the value of the ideal capacitor of the equivalent circuit could not be inferred from the complex permittivity as previously deduced, because the impedance analyzer Solartron 1260 does not allow the use

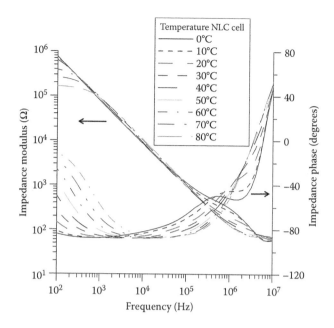

FIGURE 5.10 Experimental complex impedance of the NLC planar cell 6 as a function of the frequency for a temperature range from 0°C to 80°C.

of high voltages. Hence, the NLC cell can be electrically characterized using the experimental procedure described in Pérez et al. (2007) that is able to derive the capacitance value of the LC device as a function of the applied voltage.

Figure 5.11 shows the experimental capacitance value obtained for the NLC cell 6 applying different voltages at 5 kHz. As expected, due to the molecular reorientation of LC due to the applied electric field, the equivalent capacitance strongly depends on the applied voltage. The measured value of the capacitance ranges from 3 (when no voltage is applied and the molecules remain parallel to the substrates) to 6 nF (for applied voltages above 5 Vrms that keeps the molecules perpendicular to the substrates).

5.3.3 TEMPERATURE–FREQUENCY CONVERTER CHARACTERIZATION

The temperature–frequency converter has been implemented based on Figure 5.9b as an astable multivibrator circuit. A TL081 general purpose op-amp was used. The feedback resistors, R_1 and R_2, and the charging resistor, R, take appropriate values to set an output signal frequency of around 5 kHz. As it was deduced previously, with this working frequency, the electric behavior of the planar NLC cell can be assumed as a pure capacitance. The experimental setup implemented for the NLC square wave generator temperature sensor is shown in Figure 5.12.

A planar NLC cell used as a sensor element was placed inside a DYCOMETAL CCK-40/180 temperature-controlled chamber for accurate temperature measurements. This system was used as a temperature reference to calibrate the

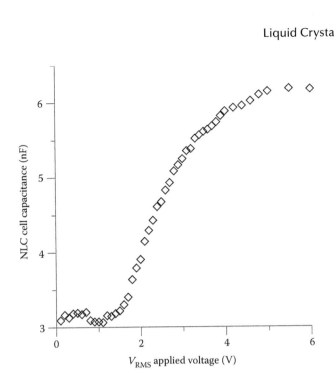

FIGURE 5.11   Equivalent capacitance of the NLC cell 6 used in the temperature–frequency converter.

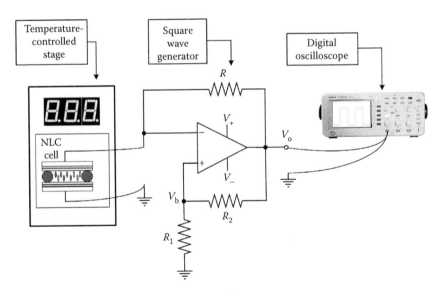

FIGURE 5.12   Experimental setup for the temperature–frequency converter.

temperature–frequency converter implemented. The astable multivibrator circuit was outside the isothermal chamber; thus, the sensor element is the only component in the circuit affected by temperature variations. Two electrical wires were used to connect the NLC cell with the circuit. The output square wave signal was measured using a digital oscilloscope.

The results obtained from the experimental setup are shown in Figure 5.13. Using the isothermal chamber, the temperature was increased linearly from room temperature (20°C) to nearly the NLC's clearing point. For this temperature range, different applied voltages to the NLC cell were tested in order to obtain the optimal value in terms of the sensitivity and linearity of the system. A starting frequency of 4.5 kHz was set at room temperature in all the tests that were carried out. Results show that the output signal's frequency is a function of the temperature. As it has been expected, as a consequence of the highest temperature dependence of the parallel permittivity ($\varepsilon_{\parallel}$), the frequency variation is greater when a higher average voltage is applied to the NLC cell.

For voltages below the threshold, the molecular orientation of LC material does not change, the effective permittivity is $\varepsilon_{\perp}$, so the output signal's frequency is almost constant with the temperature due to the small temperature dependence of $\varepsilon_{\perp}$.

When a high voltage is applied to the NLC cell, the LC molecules tend to align parallel to the direction of the electric field and perpendicular to the substrates. In this high-voltage regime, slight voltage variations have no consequence in molecular order, consequently similar curve shapes are measured for voltages above 4 Vrms.

From experimental results, the optimum values chosen for the applied voltage to the LC cell are above 4 Vrms, due to a higher variation and better linearity of the

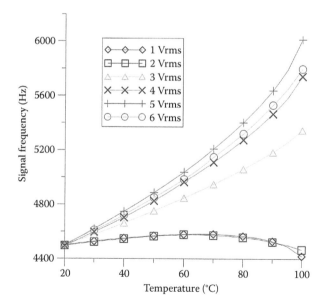

FIGURE 5.13 Variation of output frequency as a function of the temperature for different voltages applied to NLC cell.

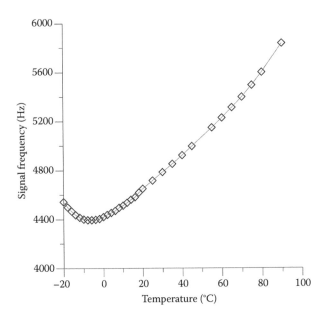

FIGURE 5.14 Output frequency variation as a function of the temperature for an NLC cell (applied voltage of 6 Vrms).

output frequency. In fact, a voltage of 6 Vrms has been applied to the NLC devices in this case. Once this voltage is fixed, the temperature range was extended to check the operating limits of the implemented system. Figure 5.14 shows the output signal frequency of the signal generator circuit as a function of the temperature. The measurements were experimentally obtained in the temperature range from −20°C to 130°C.

The sensor response can be linearly fitted in the temperature range from 0°C to 80°C with a sensitivity of 14.37 Hz/°C (Figure 5.15a).

However, in the temperature range from −6°C to 110°C, the temperature–frequency converter's response can be fitted by using a second-order polynomial function. Now, the system exhibits a lower sensitivity at low rather than at high temperatures where the sensitivity achieved is 25.2 Hz/°C (Figure 5.15b). The results demonstrate the correct operation of the temperature–frequency converter developed in a wide temperature range using an NLC cell working as an electric transducer.

5.4 LC-BASED PHASE DETECTOR AS TEMPERATURE SENSOR (TEMPERATURE-PHASE CONVERTER)

In the previous section, the design of a temperature sensor based on the temperature dependence of an NLC cell capacitance is shown. In this section, the capacitance variations with the temperature have been translated to frequency variations of a square wave signal. In particular, the capacitance variations of an LC device with the temperature will be translated to phase variations that will be converted to an output voltage change using a simple phase detector. The implemented temperature sensor is described in Figure 5.16, highlighting their main functional blocks.

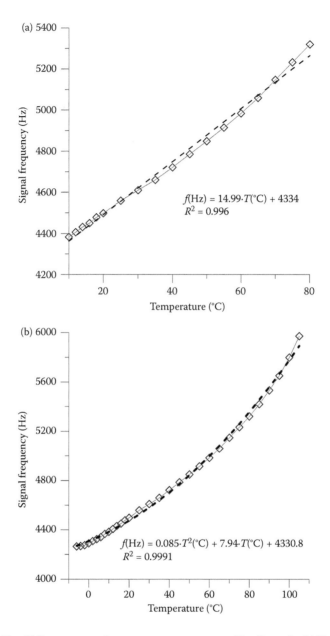

FIGURE 5.15 (a) Temperature–frequency response curve with a linear fit. (b) Temperature–frequency response curve with a second-order polynomial fit.

By comparing an input controlled sinusoidal voltage (E_s) and the voltage at the NLC cell, there is a phase shift, $\Delta\theta$, between these two waveforms that can be mathematically expressed as

$$\Delta\theta = \arctan(2 \cdot \pi \cdot f \cdot R_1 \cdot C_{CL}) - 90°, \tag{5.5}$$

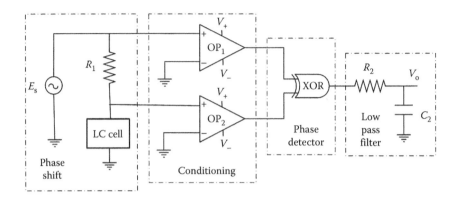

FIGURE 5.16 Electronics scheme of a phase measurement system based on an LC cell.

where R_1 is the resistance shown in Figure 5.16, C_{CL} is the equivalent capacitance of the LC cell, and f is the frequency of the applied voltage E_s. Following this equation, phase measurement can give the LC capacitance value, from which we can infer the temperature.

The temperature dependence of the capacitance of the NLC cell 4 was experimentally measured, as shown in previous sections. A variation from 8.99 to 5.43 nF when the temperature changes from 0°C to 80°C has been experimentally obtained.

The maximum sensitivity of the phase measurement system proposed is achieved using a resistance R_1 whose value satisfies the following expression

$$R_1 = \frac{1}{\left(2 \cdot \pi \cdot f \cdot \sqrt{C_{CL\,min} \cdot C_{CL\,max}}\right)}. \tag{5.6}$$

Substituting the frequency of the applied voltage ($f = 1995$ Hz corresponds to an ideal capacitive behavior of the NLC cell 4) and the extreme values of the NLC capacitance ($C_{CLmin} = 5.43$ nF and $C_{CLmax} = 8.99$ nF) at this frequency, we obtain a resistance of $R_1 = 11.41$ MΩ.

On the other hand, the relationship between the temperature, T, and the phase shift can be described by Equation 5.7.

$$\Delta\theta = -\arctan\left[\frac{\left(C_{CL\,max} - S_{CL} \cdot (T - T_0)\right)}{\sqrt{C_{CL\,min} \cdot C_{CL\,max}}}\right], \tag{5.7}$$

where the sensibility S_{CL} is $\Delta C/\Delta T = 44.5$ pF/°C which has been experimentally obtained.

Considering our experimental conditions, the phase shift takes values between $-37.85°$ and $-52.14°$. These phase shifts correspond to the extreme operating temperatures of the nematic phase of the LC, 0°C and 80°C, respectively.

The output signal of the device is voltage that is proportional to the phase shift. Several circuits can produce an output voltage proportional to the phase shift. In the proposed system (Figure 5.16), two simple comparators (OP$_1$ and OP$_2$) produce square waves from the input sinusoidal signals. These digital signals are then compared to an XOR logic gate, producing at the output a pulse train modulated by the phase shift of the initial sinusoidal signals (phase-width modulated [PWM] signal). A low-pass filter (LPF) was also added at the output of the XOR gate in order to obtain an average voltage proportional to the measured phase shift. Then, the output voltage, V_o, is given by:

$$V_o = \frac{(\Delta\theta \cdot Vcc)}{180}, \tag{5.8}$$

where Vcc is the power supply voltage of the XOR logic gate.

The LPF was designed with a cutoff frequency twenty times lower than the frequency of the pulse train (f_{PWM}). This means that the cutoff frequency has to be lower than 39 Hz. In addition, the values of the resistor and the capacitor of the filter were chosen as 16 kΩ and 1 μF, respectively. When the temperature is 0°C, the theoretical value of the output voltage is 1.18 V, which matches with the experimental one. On the other hand, when the temperature is 80°C, the theoretical output voltage is 0.85 V, while the experimental one is 0.84 V. Although they are quite similar, this mismatch may be attributed to the tolerance of the components. The comparison between experimental and theoretical results for intermediate temperatures is also shown in Figure 5.17. As it can be seen, the proposed temperature sensor operates linearly (±2%) and accurately (−4.18 mV/°C) in the considered temperature range, following the theoretical considerations.

As a summary, Table 5.1 shows the values of the electronic components used in the phase measurement scheme.

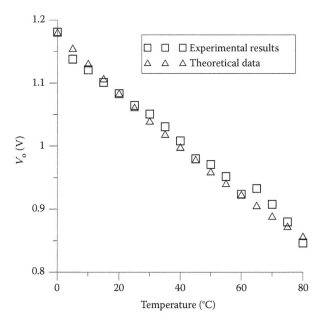

FIGURE 5.17 Experimental and theoretical output voltage of the temperature-phase converter for different temperatures in the considered range.

TABLE 5.1

Summary of the Components to Implement the Proposed Temperature Sensor

Electronic Component	Value
R_1	11.41 MΩ
R_2	16 KΩ
C_2	1 μF
XOR	7486
OP_1, OP_2	UA741

5.5 CONCLUSIONS

Unlike previously reported temperature sensors based on LC materials that need optical signals for temperature sensing, the presented system only uses electric signal to measure temperature. This work provides a new approach in the field of temperature sensing, with significant potential in applications such as *in situ* temperature sensing in embedded LC display projectors, where temperature-related malfunctions are common.

The dependence on temperature of the electrical properties of NLC cells has been analyzed, considering either a homeotropic or a planar alignment. An equivalent circuit model has been proposed and experimentally validated. As shown from our results, the capacitance of an NLC cell could be accurately used as a temperature-dependent variable.

Two important fabrication parameters, the thickness of the cell and the initial alignment, have been studied in order to understand their influence on temperature sensitivity. Theoretical results show that the sensitivity could be increased by decreasing the thickness of the cell, and we have checked this experimentally.

We propose a new electric oscillator as a temperature sensor. Temperature dependence of the NLC's permittivity causes a frequency variation in the oscillator's output signal. Different sensor sensitivities can be achieved by adjusting the control voltage applied on the NLC cell. The experimental results show that the temperature–frequency converter response can be fitted using a second-order polynomial over a wide temperature range from 6°C to 110°C. In this temperature range, the temperature–frequency converter shows good sensitivity, stability, and negligible hysteresis.

Also we proposed a new phase measurement scheme to measure the capacitance of an NLC cell. A detailed theory to describe the electric behavior of this scheme hs been developed and validated through simulations. The proposed temperature sensor operates linearly (±2%) and accurately (−4.18 mV/°C) in the considered temperature range from 0°C to 80°C.

REFERENCES

Beeckman, J., T. Hui, P.J.M. Vanbrabant et al. 2009. Polarization selective wavelength tunable filter. *Mol. Cryst. Liq. Crys.* 502: 19–28.

Blinov, L.M. and V.G. Chigrinov. 1994. *Electrooptic Effects in Liquid Crystal Materials.* New York: Springer.

Borges Neto, J.B., T.H. Silva, R.M. Assunçao et al. 2015. Sensing in the collaborative internet of things. *Sensors* 15: 6607–6632.

Costa, M.R., R.A.C. Altafim, A.P. Mammana. 2006. Electrical modeling of liquid crystal displays-LCDs. *IEEE Trans. Dielectr. Electr. Insul.* 13: 204–210.

Crossland, W.A., I.G. Manolis, M.M. Redmond et al. 2000. Holopraphic optical switching: The "roses" demonstrator. *J. Lightwave Technol.* 18: 1845–1854.

De Gennes, P.G. 1975. *The Physics of Liquid Crystal.* Oxford: Oxford University Press.

Gil-Garcia J.M., I. Garcia, J. Zubia et al. 2015. Blade tip clearance and time of arrival immediate measurement method using an optic probe. *Metrology for Aerospace IEEE* 2015: 118–122.

Hahm, J.I. and C.M. Lieber. 2004. Direct ultrasensitive electrical detection of DNA and DNA sequence variations using nanowire nanosensors. *Nano Lett.* 4: 51–54.

Kelly, S.M. and M. O'Neill. 2001. Liquid crystals for electro-optic applications. In *Handbook of Advanced Electronic and Photonic Materials and Devices*, Ten-Volume Set, ed. H.S. Nalwa. Burlingtom: Academic Press, 1–66.

Khatua, S., W.S. Chang, P. Swanglap et al. 2011. Active modulation of nanorod plasmons. *Nano Lett.* 11: 3797–3802.

Lallana, P.C., C. Vazquez, J.M.S. Pena, and R. Vergaz. 2006. Reconfigurable optical multiplexer based on liquid-crystals for polymer optical fiber networks. *Opto-Electron. Rev.* 14: 311–318.

Li, J., C. Wen, S. Gauza et al. 2005. Refractive indices of liquid crystals for display applications. *J. Display Technol.* 1: 51–61.

Marcos, C., J.M. Sánchez-Pena, J.C. Torres, and V. Urruchi. 2011. Phase-locked loop with a voltage controlled oscillator based on a liquid crystal cell as variable capacitance. *Rev. Sci. Instrum.* 82: 126101.

Perera, C., A. Zaslavsky, P. Christen et al. 2013. Sensing as a service model for smart cities supported by Internet of Things. *T. Emerg Telecommun Technol.* 25: 81–93.

Pérez I., J.M. Sánchez-Pena, J.C. Torres et al. 2007. Sinusoidal voltage-controlled Oscillator bases on a liquid crystal cell as variable capacitance. *Jpn. J. Appl. Phys.* 46: L221–L223.

Perlmutter, S.H., D. Doroski, and G. Moddel. 1996. Degradation of liquid crystal device performance due to selective adsorption of ions. *Appl. Phys. Lett.* 69: 1182.

Raistrick, I.D., D.R. Franceschetti, and J.R. Macdonald. 2005. Theory. In *Impedance Spectroscopy. Theory, Experiment and Applications*, eds. E. Barsoukov and J.R. Macdonald. New Jersey: John Wiley & Sons, 27–128.

Rashid, M.H. 1998. *Microelectronic Circuits: Analysis and Design.* Boston: PWS. pp. 1109–1112.

Si, G., Y. Zhao, E.S.P. Leong, and Y.J. Liu. 2014. Liquid-crystals-enabled active plasmonics: A review. *Materials* 7: 1296–1317.

Thurston, R.N., J. Chengm, R.B. Meyer, and G.D. Boyd. 1984. Physical mechanisms of DC switching in a liquid-crystal bistable boundary layer display. *J. Appl. Phys.* 56: 263–271.

Torres, J.C., C. Marcos, J.M. Sánchez-Pena et al. 2012. Note: series and parallel tunable resonators based on a nematic liquid crystal cell as variable capacitance. *Rev. Sci. Intrum.* 83: 086104.

Yang, D.K. and S.T. Wu. 2006. Liquid crystal materials. In *Fundamentals of Liquid Crystal Devices*, eds. D.K. Yang and S.T. Wu. Chichester: John Wiley & Sons. pp. 199–269.

Yao, K., J. Wang, X. Liu, and W. Lu. 2014. Closed-loop adaptive optics system with a single liquid crystal spatial light modulator. *Opt. Express* 22: 17216–17226.

Zhao, Q., L. Kang, B. Du et al. 2007. Electrically tunable negative permeability metamaterials based on nematic liquid crystals. *Appl. Phys. Lett.* 90: 011112.

6 Liquid Crystal-Integrated-Organic Field-Effect Transistors for Ultrasensitive Sensors

Jooyeok Seo, Myeonghun Song,
Hwajeong Kim, and Youngkyoo Kim

CONTENTS

6.1 Introduction 123
    6.1.1 LCs: *One-to-Many Stimulations* 124
    6.1.2 OFETs: *Toward Flexible Plastic Sensors* 125
6.2 Liquid Crystal-*Integrated*-Organic Field-Effect Transistors 126
    6.2.1 LC-*on*-OFET Sensory Devices 126
    6.2.2 LC-*g*-OFET Sensory Devices 129
    6.2.3 DCL-LC-*g*-OFET Sensory Devices 135
6.3 Summary and Outlook 140
Acknowledgment 141
References 141

6.1 INTRODUCTION

Advanced sensors require both high sensitivity and wide applicability for various types of environment and systems. The recent paradigm change from conventional rigid electronics to flexible electronics urges sensors to be ultrathin and flexible in order to properly fit to the flexible round-shaped systems [1]. In terms of tactile sensing, most conventional sensors have relied on either capacitive or resistive touch methods using inorganic materials and/or metallic electrodes in the presence of polymeric films [2–5]. However, these sensors need direct physical touches for sensing so that they cannot detect an indirect stimulation such as very weak airflow changes induced by approaching objects. In this regard, we have introduced liquid crystals (LCs) as a sensing component because of their excellent collective properties (group behaviors) that enable ultrasensitive sensing [6,7]. As a signal transfer and amplification part in order to benefit the feature of LCs, organic field-effect transistors (OFETs) were employed because of their potentials for ultrathin and flexible device applications [8–34]. Combining LCs and OFETs gave birth to brand-new concept devices, LC-integrated-OFET (LC-*i*-OFET) devices, which can act as a tactile sensor and many more roles in the future.

6.1.1   LCs: *One-to-Many Stimulations*

LCs, which have a directional order even in liquid states, have an intermediate phase (mesophase) between a crystalline solid state and an isotropic liquid state. As shown in Figure 6.1a, LCs can be representatively divided into six phases such as nematic, smectic, cholesteric, blue, discotic, and bowlic phases, depending on the molecular (orientational and positional) orders [35–38]. The nematic LCs have hedgehog-like topology and directional orders but no positional orders, whereas unique positional orders are present in chiral nematic (cholesteric) LCs. Some examples of LC molecules are illustrated in Figure 6.1b [39]. Such molecular orders in LCs reflect that the change of just one LC molecule upon external stimulations can significantly affect the orientations of whole or most parts of LC domains. This unique "one-to-many stimulation" motivated us to apply LCs as a sensing medium for tactile sensory devices because of its potential for achieving ultrasensitivity.

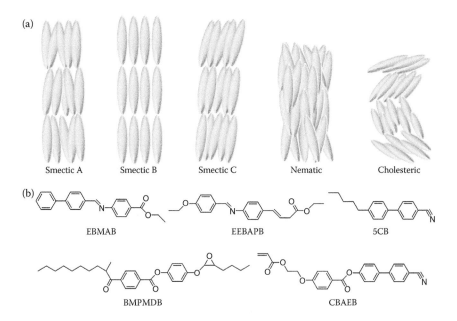

FIGURE 6.1   (a) Representative liquid crystal phases with different molecular alignment. (b) Chemical structures of representative liquid crystal molecules: (E)-ethyl-4-(biphenyl-4-yl-methyleneamino)benzoate (EBMAB—smectic A), ethyl-4-(4-((E)-4-ethoxybenzylideneamino)phenyl)but-3-enoate (EEBAPB—smectic B), 4-((3-butyloxiran-2-yl)methoxy)phenyl-4-(2-methyldecanoyl)benzoate (BMPMDB—smectic C), 4'-pentylbiphenyl-4-carbonitrile (5CB—nematic), 4'-cyanobiphenyl-4-yl 4-(2-(acryloyloxy)ethoxy)benzoate) (CBAEB—cholesteric). (Adapted from Lagerwall, J. P. F., and Scalia, G. 2012. *Curr. Appl. Phys.* 12: 1387–1412.)

6.1.2 OFETs: *Toward Flexible Plastic Sensors*

OFETs basically consist of organic channel layers, which are composed of semi-conducting organic materials including small molecules and polymers, instead of conventional inorganic channel layers [40–47]. Recently, most OFETs are fabricated by employing organic gate insulating layers that replace with inorganic insulating materials such as silicon oxides (SiOx) [48,49]. Further efforts are ongoing for polymeric electrodes, instead of conventional metallic and/or inorganic transparent conducting oxides (TCOs), in order to achieve real plastic OFETs that feature ultrathin, lightweight, and flexible shapes like a paper roll. In particular, use of wet-processable organic semiconducting materials is one of the greatest advantages for the fabrication of such flexible OFETs because low-cost manufacturing is possible at low temperatures by employing roll-to-roll processes in combination with a variety of coating technologies such as inkjet printing, slot-die coating, gravure printing, and the like [50–52].

According to the type of organic semiconducting materials, OFETs can be operated with either p-type or n-type mechanisms by applying negative or positive gate/drain voltages, respectively (Figure 6.2a). The performance of OFETs is characterized by output and transfer curves as illustrated in Figure 6.2b. The output curves

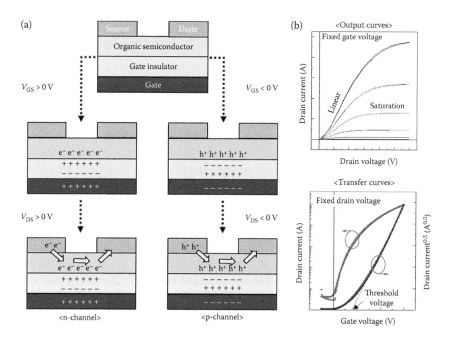

FIGURE 6.2 (a) Typical OFET structure and operating principle according to the polarity of applied drain/gate voltages: (left) positive drain/gate voltages for n-channel OFETs (n-type organic channel layers); (right) negative drain/gate voltages for p-channel OFETs (p-type organic channel layers). (b) Output (top) and transfer (bottom) curves for OFETs (note that the polarity of voltages is not given in order to express for both p-channel and n-channel OFETs).

show the change of drain current (I_{DS}) by varying the gate voltage (V_{GS}) at a fixed drain voltage (V_{DS}), which deliver information on the current saturation behavior [53–55]. On the contrary, the transfer curves provide the amplification of drain current with the gate voltage at a fixed drain voltage, which informs both on/off ratio (I_{ON}/I_{OFF}) and threshold voltage (V_{TH}) that reflect the turn-on and turn-off performances of transistor devices. The field-effect mobility, which is the value of the charge carrier drift velocity per unit electric field ($cm^2/V \cdot s$), can be calculated from the transfer curves according to following equations:

$$\text{Linear regime: } I_{DS} = \frac{W}{L}\mu C_i \left[(V_G - V_{TH})V_D - \frac{V_D^2}{2}\right] \quad (V_{DS} < V_{GS} - V_{TH}) \quad (6.1)$$

$$\text{Saturation regime: } I_{DS} = \frac{W}{2L}\mu C_i \left[(V_G - V_{TH})^2\right] \quad (V_{DS} > V_{GS} - V_{TH}) \quad (6.2)$$

6.2 LIQUID CRYSTAL-*INTEGRATED*-ORGANIC FIELD-EFFECT TRANSISTORS

Here, two different types of LC-*i*-OFETs, LC-*on*-OFETs and LC-*gate*(*g*)-OFETs, are introduced according to the geometry of OFETs. LC-*on*-OFETs consist of the LC sensing layers on top of the organic channel layers in conventional bottom gate-type OFETs, while the LC layers in LC-*g*-OFETs play a dual (sensing and gate insulating) role in the planar geometry of OFETs.

6.2.1 LC-*ON*-OFET SENSORY DEVICES

The basic concept of LC-*on*-OFET sensory devices is to bestow sensing functions to the LC layers that are placed on the organic channel layers in typical bottom-gate OFETs [56]. The fabrication process of LC-*on*-OFET sensory devices is illustrated in Figure 6.3a. First, the gate electrodes, indium-tin oxide (ITO) here, are patterned to have proper dimensions by employing various patterning methods such as photolithography/etching process for ITO-coated substrates. After the cleaning and drying steps, the gate insulating layers (poly(methyl methacrylate) [PMMA]) are coated on the ITO substrates, followed by the deposition of source/drain electrodes (silver). Next, the organic channel layers (poly(3-hexylthiophene) [P3HT]) are coated on the silver (Ag)-deposited PMMA layers on the ITO substrates. The chemical structure of materials is shown in Figure 6.3b. As observed under a scanning electron microscope (SEM) and nanoview images in Figure 6.3c, the top P3HT layers fully cover the channel area between the Ag source/drain electrodes. Finally, the LC sensing layers (4-cyano-4′-pentylbiphenyl [5CB]) are placed on top of the P3HT channel layers (Figure 6.3d). Here, it is noted that the presence of high-*k* LC molecules (dielectric constant = 18 for nematic phase) can make an induced dipole state at the interfacial regions in the P3HT channel layers.

When the LC layers in LC-*on*-OFET sensory devices are exposed to external stimulations such as very weak gas flows, at least one or more LC molecules in

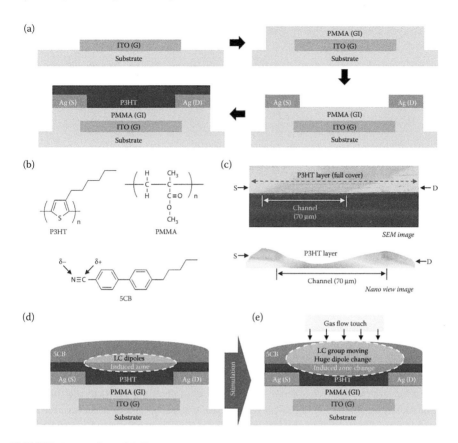

FIGURE 6.3 LC-*on*-OFET sensory devices: (a) Brief fabrication process for bottom-gate-type OFET, (b) chemical structure of materials used in this study, (c) SEM (top) and Nanoview (bottom) images for the surface of the P3HT layer on the source/drain electrodes, (d) completed LC-on-OFET sensory devices, and (e) illustration for the gas flow-stimulated state in the LC-*on*-OFET sensory devices. "S," "D," and "G" denote source, drain, and gate electrodes, respectively. (B~E: Reprinted by permission from Macmillan Publishers Ltd. *Sci. Rep.*, *Seo*, J. et al., copyright 2013.)

the LC layers are subject to change their orientations if the strength of the external stimulations is higher than the threshold energy (force) for the movement of LC molecules. In due course, the changed orientation (order) of LC molecules influences on the local dipole strength in the LC layers leading to changes in the induced dipole state at the interfacial regions in the P3HT layers that face the LC layers. This changed dipole state directly affects the channel current that flows between the source and drain electrodes by controlling gate voltages.

The bottom-gate-type OFETs fabricated showed typical p-type transistor characteristics [56]. However, as observed from the output curves in Figure 6.4a, the drain current was significantly increased when the LC layers were placed on the P3HT layers (LC-*on*-OFET devices). In addition, the shape of output curves was

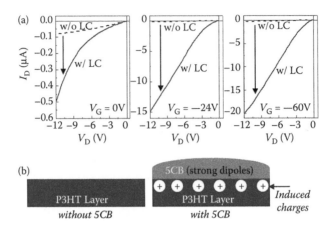

FIGURE 6.4 (a) Comparison for the output curves of the OFET devices with and without the LC layers, (b) illustration for the generation of positive charges in the channel (P3HT) layer induced by the strong dipoles of LC (5CB) molecules. (A~B: Reprinted by permission from Macmillan Publishers Ltd. *Sci. Rep., Seo*, J. et al., copyright 2013.)

largely changed without the current saturation behavior, which was similarly measured as the gate voltage increased to −60 V. This huge jump in drain current can be attributed to the generation of charges (holes) in the P3HT layers that contact the LC layers owing to the strong dipoles of LC molecules. Considering the significantly increased current in the present p-type transistor mode, it is considered that the negative dipole end of the LC molecule (5CB) might contact the surface of the P3HT layers (Figure 6.4b).

The performance of LC-*on*-OFET sensory devices was tested by varying the intensity and stimulation time of nitrogen gas flow [56]. As shown in Figure 6.5a, the drain current was gradually changed as the intensity of nitrogen flow was varied from 0.7 (11 µL/s) to 3 sccm at the fixed gas touch time (15 s). Here, it is worthy to note that 0.7 sccm is a very weak flow hence it cannot be properly felt with human skins. However, this result revealed that the response time was quite slow because of the retarded motion of LC molecules in the thick (1 mm) LC layers. As observed from Figure 6.5b, the present LC-*on*-OFET sensory devices exhibited linearly increased drain current upon stimulations of nitrogen gas flow. The sensing mechanism in the LC-*on*-OFET sensory devices can be clarified from the inset photographs in Figure 6.5a. The LC phases fluctuated upon stimulation with nitrogen flows, which was obviously measured by varying the angle of cross-polarization. Considering typical homeotropic ordering of 5CB leading to an almost dark state under 90° cross-polarization, the measured bright part in the photographs disclosed that the 5CB molecules (small domains) changed their orientation orders upon stimulation. This can be ideally illustrated, as depicted in Figure 6.5c, by taking into account the increased drain current upon stimulation (N_2 ON state).

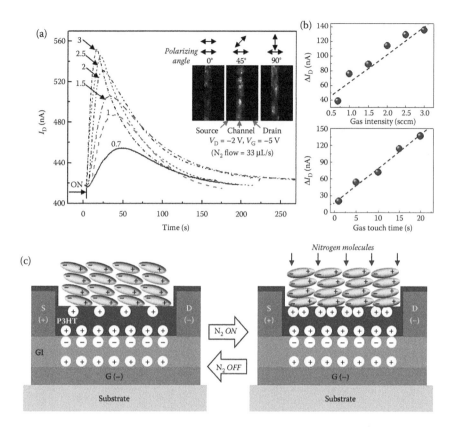

FIGURE 6.5 Sensing performances of LC-*on*-OFET sensory devices and mechanism: (a) Drain current (I_D) as a function of stimulation time according to the intensity (strength) of nitrogen gas flow (sccm). (b) Drain current difference as a function of nitrogen gas intensity (stimulation time = 15 s) (top panel) and nitrogen gas stimulation time (nitrogen gas intensity = 2 sccm) (bottom panel). (c) Illustration for the orientation change of 5CB molecules in the LC layers before and after the nitrogen gas stimulation. (A~C: Reprinted by permission from Macmillan Publishers Ltd. *Sci. Rep.*, Seo, J. et al., copyright 2013.)

6.2.2 LC-*G*-OFET Sensory Devices

To further extend the concept of LC-*on*-OFET devices, a planar-type structure has been introduced by changing the location (position) of electrodes. As shown in Figure 6.6a, the source/drain and gate electrodes are aligned on the same plane of substrates [57]. The P3HT channel layers were coated on the regions between the source and drain electrodes, whereas any remaining P3HT layers between the source and gate electrodes were removed in order to prevent possible leakage paths. Next, the LC layers were placed on the whole part between the drain and gate electrodes, leading to a planar-type OFET (i.e., LC-*g*-OFET) that is operated by the gate insulating role of LC.

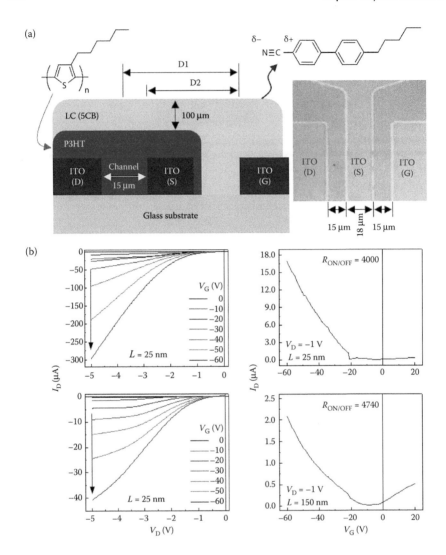

FIGURE 6.6 (a) Illustration for the device structure of LC-g-OFET and the chemical structure of 5CB and P3HT (right: optical microscope image for the patterned ITO-glass substrate). (b) Output (left) and transfer (right) curves of the LC-g-OFET devices according to the P3HT thickness (top: 25 nm, bottom: 150 nm). (A~B: Reprinted with permission from Seo, J. et al. 2015. Liquid crystal-gatedorganic field-effect transistors with in-plane drain–source–gate electrode structure. *ACS Appl. Mater.* Interfaces 7: 504–510. Copyright 2015 American Chemical Society.)

As shown in Figure 6.6b (left), the output curves of LC-g-OFET were sensitive to the thickness of the P3HT channel layers. No drain current saturation was measured for the devices with the thin (25 nm) P3HT layers, while the devices with the thick (150 nm) P3HT layers exhibited noticeable saturation behaviors in the output curves. This result implies that the strong dipole moment of LC molecules significantly

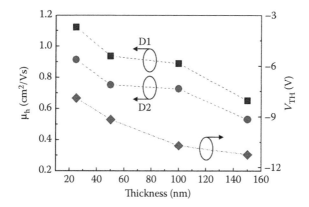

FIGURE 6.7 Hole mobility (μ_h) and threshold voltage (V_{TH}) for the LC-*g*-OFET devices as a function of P3HT thickness. Two different dimensions (see D1 and D2 in Figure 6.6a) were used to calculate the capacitance (C_i) of the 5CB gate insulator. (Reprinted with permission from Seo, J. et al. 2015. Liquid crystal-gatedorganic field-effect transistors with in-plane drain–source–gate electrode structure. *ACS Appl. Mater. Interfaces* 7: 504–510. Copyright 2015 American Chemical Society.)

influences on the transistor characteristics of LC-*g*-OFET devices. Nevertheless, both devices showed quite nice transfer curves at very low drain voltage (−1 V) with reasonable on/off ratios (>4000) (Figure 6.6b [right]). The hole mobility and threshold voltage of devices also varied with the P3HT thickness, which might be related with the channel volume affected by the LC layers (Figure 6.7). The resulting hole mobility was ca. 0.5~1.2 cm²/Vs, which is one of the highest hole mobilities reported for the OFETs with the P3HT layers. However, it is noted that the threshold voltage is relatively high compared with the trend of drain voltage. This can be attributable to the long distance between the channel region and the gate electrode (see D1 = 40.5 μm), which is a large value compared to the typical thickness of gate insulating layers (300~700 nm). Hence, the threshold (turn-on) voltage (i.e., V_{TH}) is expected to be further reduced through the controlled distance between the channel region and the gate electrode.

The operation mechanism of LC-*g*-OFET devices was investigated by employing a polarized optical microscope that can measure the orientation orders of LC molecules in the channel regions. As shown in Figure 6.8a (top), the bright and dark images were measured under linear and cross-polarization conditions, respectively, at no bias condition. Additional studies with various polarization angles revealed that the 5CB molecules made a tilted homeotropic alignment in this case. When applying slightly low gate (−15 V) and drain (−1 V) voltages, both the channel area and the S–G interelectrode area became dark but not perfectly dark under cross-polarization, supporting insufficiently controlled (flipped) 5CB dipoles.

When the voltages were further increased, a perfectly dark image was observed for the channel area under cross-polarization (note that the surrounding part was still dark gray). Note that the dipole direction in the channel (D–S interelectrode) area was flipped from (− to +) at $V_D = -1$ V and $V_G = -15$ V (see the middle images

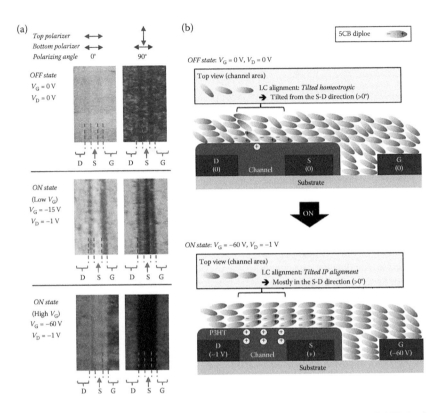

FIGURE 6.8 **(See color insert.)** (a) Optical microscope images for the LC-*g*-OFET devices according to the drain and gate voltages: (left) linear polarization state (0°), (right) cross-polarization state (90°). (b) Illustration for the movement of LC molecules according to the drain and gate voltages: (top) OFF state, (bottom) ON state. (A~B: Reprinted with permission from Seo, J. et al. 2015. Liquid crystal-gatedorganic field-effect transistors with in-plane drain–source–gate electrode structure. *ACS Appl. Mater. Interfaces* 7: 504–510. Copyright 2015 American Chemical Society.)

in Figure 6.8a) to (+ to −) at $V_D = -1$ V and $V_G = -60$ V (see the bottom images in Figure 6.8a) by the strong gate voltage (−60 V) that overwhelms the weak drain voltage (−1 V). This result indicates that 5CB molecules in the channel area were aligned parallel to the direction of D–S–G electrodes under a sufficiently high gate bias condition (Figure 6.8b). However, it is considered that the presence of weak drain voltage (−1 V) with the counterpole against gate voltage will affect the LC alignment to be slightly tilted from the surface of the P3HT layer (see the channel area in Figure 6.8b). As a consequence, the negative dipole end of the 5CB molecules induces charges (holes) on the interfacial area of the P3HT layer. In addition, due to the high dipole feature of LC (5CB), the strength of gate voltages can make more dielectric polarization through the LC layer, as always observed in the case of conventional gate dielectric (insulating) layers in normal OFET structures. Therefore, it is clear that the LC molecules play a gate insulating role in the planar OFET structures, leading to LC-*g*-OFET devices.

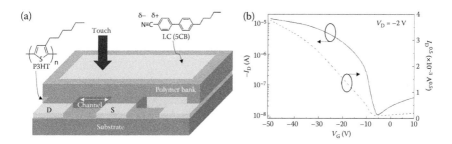

FIGURE 6.9 (a) Illustration for the device structure of the LC-*g*-OFET device with the silicone polymer bank. Note that P3HT and 5CB were used as a channel layer and a sensing gate insulator layer, respectively. (b) Transfer curves for the present LC-*g*-OFET device at $V_D = -2$ V. (A~B: Reprinted with permission from Seo, J. et al. 2014. Touch sensors based on planar liquid crystal-gated-organic field-effect transistors. *AIP Adv.* 4: 097109. Copyright 2014, American Institute of Physics.)

Next, the LC-*g*-OFET devices were examined as a tactile sensor after optimization of device performances. As shown in Figure 6.9a, a silicone polymer bank was attached to the LC-*g*-OFET devices in order to hold LC molecules during the touch events [58]. The transfer curve showed that the drain current (I_D) of the LC-*g*-OFET device with the silicone bank was gradually increased up to >10 μA as the gate voltage was increased to −50 V at the drain voltage of −2 V (Figure 6.9b). The hole mobility of the optimized LC-*g*-OFET device reached ~1.5 cm²/Vs, which is comparable to the charge mobility of inorganic FETs with amorphous silicon channel layers. On the basis of this excellent performance of LC-*g*-OFET, the stimulation test was carried out by employing a nitrogen gas flow.

As shown in Figure 6.10a, the drain current changed very slightly by the stimulation with the nitrogen flow (5 sccm for 5 s) in the case of no gate bias condition at $V_G = 0$ V and $V_D = -0.1$ V. Although this result informs the possible signal change

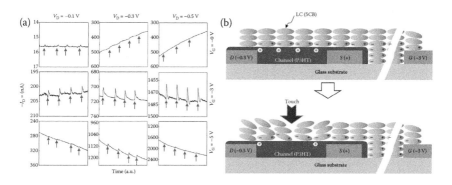

FIGURE 6.10 (a) Drain current (I_D) change upon stimulation with nitrogen gas flow (5 sccm for 5 s) in the LC-*g*-OFET device. (b) Possible mechanism for the negative drain current upon the stimulation event for the present LC-*g*-OFET devices. (A~B: Reprinted with permission from Seo, J. et al. 2014. Touch sensors based on planar liquid crystal-gated-organic field-effect transistors. *AIP Adv.* 4: 097109. Copyright 2014, American Institute of Physics.)

upon external stimulation at no gate bias condition, it rather confirms that the drain current signal is too weak to be practically applied without any gate bias. When the gate voltage was increased up to $V_G = -3$ V, the drain current signal had dramatically changed upon nitrogen flow stimulation. When the gate voltage was further increased to -5 V, the drain current signal became weaker again irrespective of the drain voltages. This result implies the presence of an optimum voltage condition for the best signal in the present LC-g-OFET devices. Here it is worthy to note that the drain current was reduced upon nitrogen flow stimulation. This can be attributed to the applied voltage conditions that are lower than the threshold (gate) voltage (see the transfer curve in Figure 6.9b). As illustrated in Figure 6.10b (top), the 5CB molecules can make a parallel alignment with the direction of electric field but they were weakly forced at the low gate voltages. So the nitrogen flow could easily change the movement of 5CB molecules so as to make random orientations. During the stimulation, the negative dipoles of 5CB molecules might flip from the direction of the

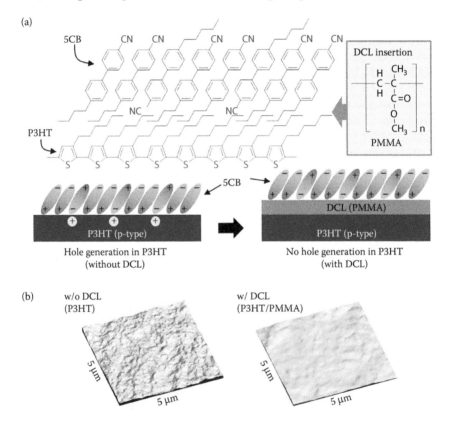

FIGURE 6.11 (a) Concept for DCL (PMMA) to block the charge generation in the channel (P3HT) layer. (b) AFM images (height mode) for the surface of the P3HT layer (left) and the DCL (PMMA)-coated P3HT layer (right). (A~B: Seo, J. et al. 2015. Ultrasensitive tactile sensors based on planar liquid crystal-gated-organic field-effect transistors with polymeric dipole control layers. *RSC Adv.* 5: 56904. Reproduced by permission of The Royal Society of Chemistry.)

P3HT layers so that it could reduce the amount of charges (holes) in the P3HT layer. This might eventually result in decreased drain current upon nitrogen gas stimulation. Considering the sensing characteristics of LC-*g*-OFET devices, the generation of charges in the P3HT layers by the presence of LC molecules delivers an adverse effect on the device performances including high leakage currents, etc.

6.2.3 DCL-LC-*G*-OFET SENSORY DEVICES

In order to minimize the inherent leakage current in the LC-*g*-OFET devices owing to the charge generation caused by the presence of high-*k* LC molecules, a low-*k* interlayer, so-called dipole control layer (DCL), has been introduced between the P3HT channel layer and the LC layer [59]. As depicted in Figure 6.11a, the insertion of DCL can block the strong influence of high-*k* 5CB molecules on the P3HT layer. When the 10-nm-thick PMMA layer was used as a DCL, the surface morphology was noticeably changed from a rough phase to a smoother phase (Figure 6.11b). In addition, the contact angle of the DCL-coated P3HT layer was pronouncedly reduced as the result of the changed surface morphology [59].

As shown from the current–voltage (I–V) curves in Figure 6.12a, the drain current of planar diode devices was pronouncedly reduced by the insertion of the DCL layer between the P3HT layer and the LC layer. The reduction in drain current was >100-fold by the presence of only the 10-nm-thick PMMA layer (DCL) whose dielectric constant is ~3.5. The same PMMA DCL was introduced for the fabrication of LC-*g*-OFET devices, which led to DCL-LC-*g*-OFET devices. As shown in Figure 6.12b, the transfer curve of DCL-LC-*g*-OFET devices was greatly improved in terms of on/

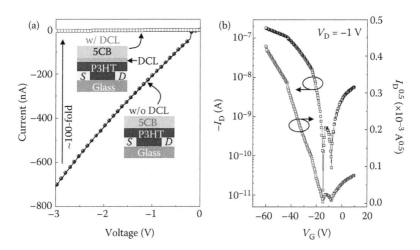

FIGURE 6.12 (a) Current–voltage (I–V) characteristics for the planar-type diode devices with and without the 10-nm-thick PMMA DCL (inset: device structures). (B) Transfer curve for the planar DCL-LC-g-OFET device at the drain voltage of −1 V. (A~B: Seo, J. et al. 2015. Ultrasensitive tactile sensors based on planar liquid crystal-gated-organic field-effect transistors with polymeric dipole control layers. *RSC Adv.* 5: 56904. Reproduced by permission of The Royal Society of Chemistry.)

off ratio compared to that of LC-*g*-OFET devices (Figure 6.9b). This result supports that the presence of the 10-nm-thick DLC contributes to the reduction of leakage current as well as the improved transistor performance.

The sensing performance of the DCL-LC-*g*-OFET devices was examined by stimulating the channel region with nitrogen gas flows. As shown in Figure 6.13a (top), the drain current gradually increased as the intensity of nitrogen flow increased from 0.5 to 4 sccm at the same stimulation time (20 s). A maximum value was apparently measured at around 20 s, which is in accordance with the application time of the nitrogen flow. When the nitrogen flow was switched off, the drain current slowly decreased with time (up to ~80 s) for all cases. As discussed for the LC-*on*-OFET devices, this slow behavior can be still attributable to the dynamic motion of LC molecules in the 100-μm-thick LC layer. It is noted that the sensing range of the DCL-LC-*g*-OFET devices extended down to 0.5 sccm (~8.3 μL/s) compared with

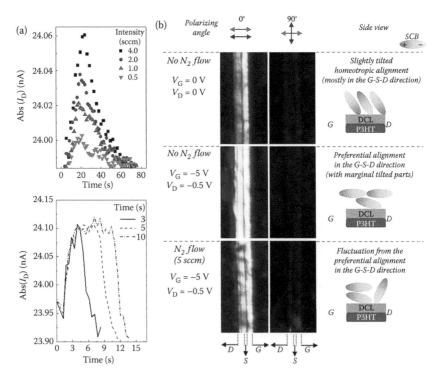

FIGURE 6.13 (a) Change of drain current with time upon stimulation of nitrogen gas flows by varying (top) the nitrogen gas intensity from 0.5 to 4.0 sccm for 20 s each and (bottom) the stimulation time of the nitrogen gas at a fixed intensity of 5 sccm. (b) Comparison of optical microscope images on the channel area in the DCL-LC-*g*-OFET devices and illustration of corresponding LC alignment: The drain and gate voltages are given on the left of each image, while the polarization angle is displayed on top. (A~B: Seo, J. et al. 2015. Ultrasensitive tactile sensors based on planar liquid crystal-gated-organic field-effect transistors with polymeric dipole control layers. *RSC Adv.* 5: 56904. Reproduced by permission of The Royal Society of Chemistry.)

0.7 sccm by the LC-*on*-OFET devices (Figure 6.5a). As expected, the maximum point of the drain current signal shifted toward a longer time upon increasing the stimulation time of nitrogen gas flow (Figure 6.13a [bottom]).

The sensing mechanism of DCL-LC-*g*-OFET devices was investigated with optical microscopy by changing the applied voltages under linear and/or cross-polarization conditions. As shown in Figure 6.13b, the brightness in the channel area under linear polarization (0°) was noticeably increased by increasing the gate (−5 V) and drain (−0.5 V) voltages before stimulating with the nitrogen gas flow. The image under cross-polarization (90°) became much darker at the biased condition ($V_G = -5$ V and $V_D = -0.5$ V). This result indicates that the LC orientation was changed from the tilted homeotropic alignment to the preferential planar alignment in the direction of drain–source–gate bias. When the nitrogen gas was applied to the

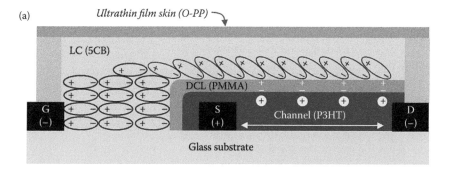

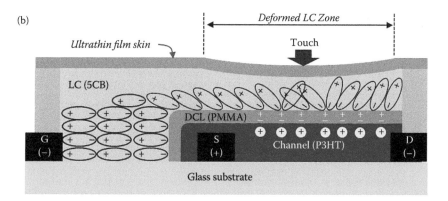

FIGURE 6.14 (a) Illustration for the fluctuation of the LC layer in the planar DCL-LC-*g*-OFET device upon stimulation of external touch (nitrogen flow) at $V_G = -5$ V and $V_D = -0.5$ V. (b) Drain current as a function of time upon stimulation of nitrogen flows at various intensities (note that the stimulation time was 20 s). (c) Drain current as a function of time upon stimulation of nitrogen flow (5 sccm) by varying the stimulation time from 3 to 20 s (the total duration time (t_D) is marked for example). (A~B: Seo, J. et al. 2015. Ultrasensitive tactile sensors based on planar liquid crystal-gated-organic field-effect transistors with polymeric dipole control layers. *RSC Adv.* 5: 56904. Reproduced by permission of The Royal Society of Chemistry.)

channel area in the DCL-LC-*g*-OFET device at the same biased condition (ON state), a fluctuating brightness was measured in the channel region under a linear polarization condition. In particular, the cross-polarization image disclosed that there is a bright part in the channel area even though the surrounding regions are all black. This result supports that the 5CB molecules were subject to significant change in orientation by the stimulation of nitrogen gas flow. On the basis of the fluctuated LC orientation (alignment) upon the nitrogen gas stimulation, the sensing mechanism in the DCL-LC-*g*-OFET devices can be illustrated as shown in Figure 6.14. Here, it is noted that the presence of the ultrathin skin (polymer film) might reduce the device's sensitivity but it could also act as a protection layer for practical applications.

Finally, the DCL-LC-*g*-OFET devices were examined as a direct physical touch sensor, not by gas flow but by a pencil or finger, as illustrated in Figure 6.15a [60]. The optimized DCL-LC-*g*-OFET devices showed typical p-type transistor characteristics with the on/off ratio of $\sim 10^4$ (Figure 6.15b). In particular, the present devices

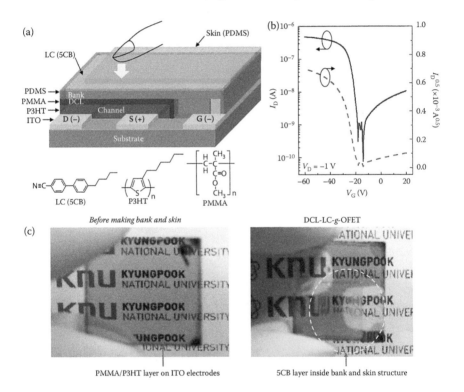

FIGURE 6.15 (**See color insert.**) (a) Illustration for the device structure of DCL-LC-*g*-OFET device with the poly(dimethyl siloxane) (PDMS) protection skin layer: The width and length of ITO electrodes (D, S, and G) were 18 μm and 3 mm, respectively. (b) Transfer characteristics of the DCL-LC-*g*-OFET device at $V_D = -1$ V. (c) Photographs for the DCL-LC-*g*-OFET device with (left) and without (right) the PDMS bank and skin parts. (A~C: Reprinted from *Org. Electron.*, 28, Seo, J. et al., Physical force-sensitive touch responses in liquid crystal-gated-organic field-effect transistors with polymer dipole control layers., 184–188, Copyright 2016, with permission from Elsevier.)

with the PDMS skin layer exhibited clear increase/saturation behaviors in drain current with the gate voltage at a fixed drain voltage (-1 V). However, it is noted that the threshold voltage is still high compared to the drain voltage, which should be further optimized. As observed from the photographs in Figure 6.15c, the present DCL-LC-g-OFET devices can be used as a semitransparent physical sensor.

As shown in Figure 6.16a, the drain current quickly increased upon physical touch by a pencil-like load (1.2 g) at $V_G = -20$ V and $V_D = -1$ V [60]. The repeated test showed that almost similar drain current values with a narrow standard deviation were measured for each touch event (Figure 6.16b). The touch test was extended to various loads from 0.6 to 4.8 g by fixing the applied voltages ($V_G = -20$ V and $V_D = -1$ V). As shown in Figure 6.16c, the drain current gradually increased as the load was increased. This result could be well repeated when different devices were used for the touch experiment [60]. The response time measured from the rise and decay curve in Figure 6.16d here was as quick as less than 300 ms, even though the time resolution of the present measurement system is $\sim$150 ms. This result reflects that more fine rise/decay responses can be measured if new measurement systems with a high time resolution are employed.

As shown in Figure 6.17a, the drain current change upon stimulation was almost linearly proportional to the strength (load) of physical touches. Interestingly, the rise/decay time also linearly increased with the strength of physical touches, which

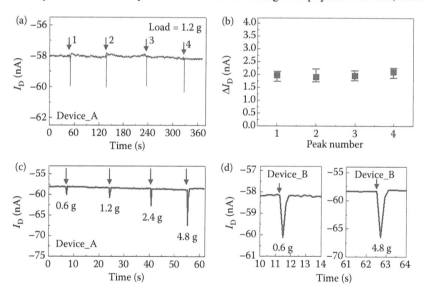

FIGURE 6.16 (a) Repeated drain current responses upon physical touches (1~4) (load = 1.2 g [0.38 N/cm²]) to the channel region in the DCL-LC-g-OFET devices. (b) Deviation of drain current signals upon repeated physical touches. (c) Drain current responses upon physical touches with different touch loads (0.6 g [0.19 N/cm²]~4.8 g [1.52 N/cm²]). (d) Drain current peaks (left: 0.6 g; right: 4.8 g) from Device_B of which response is very similar to Device_A. (A~D: Reprinted from *Org. Electron.*, 28, Seo, J. et al., Physical force-sensitive touch responses in liquid crystal-gated-organic field-effect transistors with polymer dipole control layers., 184-188, Copyright 2016, with permission from Elsevier.)

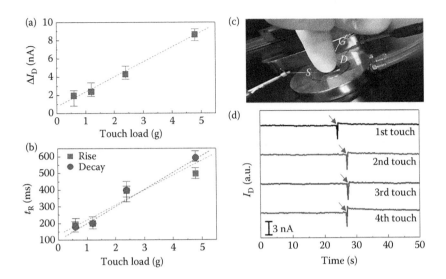

FIGURE 6.17 (a, b) Change of drain current (a) and response time (t_R) (b) as a function of physical touch load (strength) in the DCL-LC-g-OFET devices. (c) Photograph for the direct physical touch with a human finger to the channel layer in the DCL-LC-g-OFET device. (d) Change of drain current (I_D) response by the direct physical touch with human fingers on the channel area in the DCL-LC-g-OFET device ($V_D = -1$ V). (A~D: Reprinted from *Org. Electron.*, 28, Seo, J. et al., Physical force-sensitive touch responses in liquid crystal-gated-organic field-effect transistors with polymer dipole control layers., 184–188, Copyright 2016, with permission from Elsevier.)

may be related with the response time of 5CB molecules because the number of 5CB molecules affected by stimulation could be different with the strength of physical touches. In other words, the different response time could be very useful to get additional information on the touch strength beside the drain current value. It is also worthy to note that the response time is less than ~600 ms even by the heaviest load (4.8 g) applied in this work. When a human finger touched the channel part in the DCL-LC-g-OFET devices (Figure 6.17c), the drain current quickly changed as shown in Figure 6.17d. Although a strong touch was repeatedly applied by human fingers, the present DCL-LC-g-OFET devices exhibited stable drain current changes.

6.3 SUMMARY AND OUTLOOK

The combination of LC and OFET opened new category of sensory devices such as LC-*on*-OFET, LC-g-OFET, and DCL-LC-g-OFET. The LC-*on*-OFET devices possess huge potentials in sensing ultralow-level gas flows, which can be effectively applied for various high sensitivity detection systems such as humanoid robots, surveillance systems, and military defense systems. In addition, the LC-g-OFET and DCL-LC-g-OFET devices are expected to be applied for low-cost sensor systems because of their simple and cost-effective fabrication processes. Finally, both sensory devices can pave a bright way to achieve efficient and low-cost biomedical sensors via tailoring and functionalizing the LC materials to diagnose various diseases.

ACKNOWLEDGMENT

This work was financially supported by the grants from the Korean Government (Human Resource Training Project for Regional Innovation_MOE_NRF-2014H1C1A1066748, Basic Science Research Program_2009-0093819, NRF_2015R1A2A2A01003743, NRF_2014R1A1A3051165, NRF_2016H1D5A1910319, NRF_2014H1A2A1016454).

REFERENCES

1. Russo, A., Ahn, B., Adams, J. J., Duoss, E. B., Bernhard, J. T., and Lewis, J. A. 2011. Pen-on-paper flexible electronics. *Adv. Mater.* 23: 3426–3430.
2. Ko, W. and Wang, Q. 1999. Touch mode capacitive pressure sensors. *Sens. Actuators* 75: 242–251.
3. Pan, L., Chortos, A., Yu, G., Wang, Y., Isaacson, S., Allen, R., Shi, Y., Dauskardt, R., and Bao, Z. 2014. An ultra-sensitive resistive pressure sensor based on hollow-sphere microstructure induced elasticity in conducting polymer film. *Nat. Commun.* 5: 3002.
4. Nam, S., Seo, J., Park, S., Lee, S., Jeong, J., Lee, H., Kim, H., and Kim, Y. 2013. Hybrid phototransistors based on bulk heterojunction films of poly(3-hexylthiophene) and zinc oxide nanoparticle. *ACS Appl. Mater. Interfaces* 5: 1385.
5. Park, S., Nam, S., Kim, J., Seo, J., Jeong, J., Woo, S., Kim, H., and Kim, Y. 2013. Influence of nickel(II) oxide nanoparticle addition on the performance of organic field effect transistors. *J. Nanosci. Nanotechnol.* 13: 6016.
6. Collings, P. J. and Hird, M. 1997. *Introduction to Liquid Crystals: Chemistry and Physics.* Boca Raton: CRC Press, pp. 1–298.
7. Blinov, L. M. 2011. *Structure and Properties of Liquid Crystals.* New York: Springer, pp. 1–439.
8. Kymissis, I. 2009. *Organic Field Effect Transistors: Theory, Fabrication and Characterization.* New York: Springer, pp. 1–146.
9. Kim, C. H., Bonnassieux, Y., and Horowitz, G. 2014. Compact DC modeling of organic field-effect transistors: Review and perspectives. *IEEE Trans. Electron Devices* 61: 278–287.
10. Zhang, C., Chen, P., and Hu, W. 2015. Organic field-effect transistor-based gas sensors. *Chem. Soc. Rev.* 44: 2087–2107.
11. Zang, Y., Zhang, F., Huang, D., Gao, X., Di, C., and Zhu, D. 2015. Flexible suspended gate organic thin-film transistors for ultra-sensitive pressure detection. *Nat. Commun.* 6: 6269.
12. Welch, M. E., Doublet, T., Bernard, C., Malliaras, G. G., and Ober, C. K. 2015. A glucose sensor via stable immobilization of the GOx enzyme on an organic transistor using a polymer brush. *J. Polym. Sci., Part A: Polym. Chem.* 53: 372–377.
13. Mirza, M., Wang, J., Wang, L., He, J., and Jiang, C. 2015. Response enhancement mechanism of NO_2 gas sensing in ultrathin pentacene field-effect transistors. *Org. Electron.* 24: 96–100.
14. Crone, B., Dodabalapur, A., Gelperin, A., Torsi, L., Katz, H. E., Lovinger, A. J., and Bao, Z. 2001. Electronic sensing of vapors with organic transistors. *Appl. Phys. Lett.* 78: 2229.
15. Das, A., Dost, R., Richardson, T., Grell, M., Morrison, J. J., and Turner, M. L. 2007. A nitrogen dioxide sensor based on an organic transistor constructed from amorphous semiconducting polymers. *Adv. Mater.* 19: 4018–4023.
16. Torsi, L., Dodabalapur, A., Sabbatini, L., and Zambonin, P.G. 2000. Multi-parameter gas sensors based on organic thin-film-transistors. *Sens. Actuators, B* 67: 312–316.
17. Zhu, Z. T., Mason, J. T., Dieckmann, R., and Malliaras, G. G. 2002. Humidity sensors based on pentacene thin-film transistors. *Appl. Phys. Lett.* 81: 4643.

18. Roberts, M. E., Sokolov, A. N., and Bao, Z. 2009. Material and device considerations for organic thin-film transistor sensors. *J. Mater. Chem.* 19: 3351–3363.
19. Darlinski, G., Böttger, U., Waser, R., Klauk, H., Halik, M., Zschieschang, U., Schmid, G., and Dehm, C. 2005. Mechanical force sensors using organic thin-film transistors. *J. Appl. Phys.* 97: 093708.
20. Torsi, L., Magliulo, M., Manoli, K., and Palazzo, G. 2013. Organic field-effect transistor sensors: A tutorial review. *Chem. Soc. Rev.* 42: 8612–8628.
21. Andringa, A. M., Spijkman, M. J., Smits, E. C. P., Mathijssen, S. G. J., Setayesh, S., Willard, N. P., Borshchev, O. V., Ponomarenko, S. A., Blom, P. W.M., and Leeuw, D. M. D. 2010. Gas sensing with self-assembled monolayer field-effect transistors. *Org. Electron.* 11: 895–898.
22. Jeong, J., Lee, Y., Kim, Y., Park, Y., Choi, J., Park, T., Soo, C., Won, S., Han, I., and Ju, B. 2010. The response characteristics of a gas sensor based on poly-3-hexylithiophene thin-film transistors. *Sens. Actuators, B* 146: 40–45.
23. Lin, P. and Yan, F. 2012. Organic thin-film transistors for chemical and biological sensing. *Adv. Mater.* 24: 34–51.
24. Someya, T., Dodabalapur, A., Huang, J., See, K. C., and Katz, H. E. 2010. Chemical and physical sensing by organic field-effect transistors and related devices. *Adv. Mater.* 22: 3799–3811.
25. Lee, X., Sugawara, Y., Ito, A., Oikawa, S., Kawasaki, N., Kaji, Y., Mitsuhashi, R. et al. 2010. Quantitative analysis of O_2 gas sensing characteristics of picene thin film field-effect transistors. *Org. Electron.* 11: 1394–1398.
26. Jeong, S., Jeong, J., Chang, S. Kang, S. Cho, K., and Ju, B. 2010. The vertically stacked organic sensor-transistor on a flexible substrate. *Appl. Phys. Lett.* 97: 253309.
27. Dutta, S. and Dodabalapur, A. 2009. Zinc tin oxide thin film transistor sensor. *Sens. Actuators, B* 143: 50–55.
28. Dahiya, R. S., Metta, G., Valle, M., Adami, A., and Lorenzelli, L. 2009. Piezoelectric oxide semiconductor field effect transistor touch sensing devices. *Appl. Phys. Lett.* 95: 034105.
29. Li, B. and Lambeth, D. N. 2008. Chemical sensing using nanostructured polythiophene transistors. *Nano Lett.* 8: 3563–3567.
30. Manunza, I. and Bonfiglio, A. 2007. Pressure sensing using a completely flexible organic transistor. *Biosens. Bioelectron.* 22: 2775–2779.
31. Bai, H. and Shi, G. 2007. Gas sensors based on conducting polymers. *Sensors* 7: 267–307.
32. Manunza, I., Sulis, A., and Bonfiglio, A. 2006. Pressure sensing by flexible, organic, field effect transistors. *Appl. Phys. Lett.* 89: 143502.
33. Liao, F., Chen, C., and Subramanian, V. 2005. Organic TFTs as gas sensors for electronic nose applications. *Sens. Actuators, B* 107: 849–855.
34. Fukuda, H., Ise, M., Kogure, T., and Takano, N. 2004. Gas sensors based on poly-3-hexylthiophene thin-film transistors. *Thin Solid Films* 464–465: 441–444.
35. Yu, L. J. and Saupe, A. 1980. Observation of a biaxial nematic phase in potassium laurate-1-decanol-water mixtures. *Phys. Rev. Lett.* 45: 1000.
36. Merkel, K., Kocot, A., Vij, J. K., Korlacki, R., Mehl, G. H., and Meyer, T. 2004. Thermotropic biaxial nematic phase in liquid crystalline organo-siloxane tetrapodes. *Phys. Rev. Lett.* 93: 237801.
37. Gim, M.-J., Turlapati, S., Debnath, S., Rao, N. V. S., and Yoon, D. 2016. Highly polarized fluorescent illumination using liquid crystal phase. *ACS Appl. Mater. Interfaces* 8: 3143–3149.
38. Jansze, S. M., Martínez-Felipe, A., Storey, J. M. D., Marcelis, A. T. M., and Imrie, C. T. 2015. A twist-bend nematic phase driven by hydrogen bonding. *Angew. Chem. Int. Ed.* 54: 643–646.

39. Lagerwall, J. P. F. and Scalia, G. 2012. A new era for liquid crystal research: Applications of liquid crystals in soft matter nano-, bio- and microtechnology. *Curr. Appl. Phys.* 12: 1387–1412.

40. Muccini, M. 2006. A bright future for organic field-effect transistors. *Nat. Mater.* 5: 605–613.

41. Yamashita, Y. 2009. Organic semiconductors for organic field-effect transistors. *Sci. Technol. Adv. Mater.* 10: 024313.

42. Kergoat, L., Herlogsson, L., Braga, D., Piro, B., Pham, M.-C., Crispin, X., Berggren, M., and Horowitz, G. 2010. A water-gate organic field-effect transistor. *Adv. Mater.* 22: 2565–2569.

43. Braga, D., Erickson, N. C., Renn, M. J., Holmes, R. J., and Daniel Frisbie, C. 2012. High-transconductance organic thin-film electrochemical transistors for driving low-voltage red-green-blue active matrix organic light-emitting devices. *Adv. Funct. Mater.* 22: 1623–1631.

44. Cho, J. H., Lee, J., Xia, Y., Kim, B., He, Y., Renn, M. J., Lodge, T. P., and Frisbie, C. D. 2008. Printable ion-gel gate dielectrics for low-voltage polymer thin-film transistors on plastic. *Nat. Mater.* 7: 900.

45. Kim, Y. and Ha, C. S. 2008. *Advances in Organic Light-Emitting Device.* Zuerich: Trans Tech Publications.

46. Kim, H., Nam, S., Jeong, J., Lee, S., Seo, J., Han, H., and Kim, Y. 2014. Organic solar cells based on conjugated polymers: History and recent advances. *Korean J. Chem. Eng.* 31: 1095–1104.

47. Nam, S., Kim, J., Lee, H., Kim, H., Ha, C.-S., and Kim, Y. 2012. Doping effect of organosulfonic acid in poly (3-hexylthiophene) films for organic field-effect transistors. *ACS Appl. Mater. Interfaces* 4: 1281.

48. Yoon, M.-H., Kim, C., Facchetti, A., and Marks, T. J. 2006. Gate dielectric chemical structure–organic field-effect transistor performance correlations for electron, hole, and ambipolar organic semiconductors. *J. Am. Chem. Soc.* 128: 12851–12869.

49. Veres, J., Ogier, S., Lloyd, G., and Leeuw, D. M. D. 2004. Gate insulators in organic field-effect transistors. *Chem. Mater.* 16: 4543–4555.

50. Bao, Z., Dodabalapur A., and Lovinger, A. 1996. Soluble and processable regioregular poly(3-hexylthiophene) for thin film field-effect transistor applications with high mobility. *Appl. Phys. Lett.* 69: 4108.

51. Sirringhaus, H. Device physics of solution-processed organic field-effect transistors. *Adv. Mater.* 17: 2411.

52. Nam, S., Lee, S., Lee, I., Shin, M., Kim, H., and Kim, Y. 2011. Nanomorphology-driven two-stage hole mobility in blend films of regioregular and regiorandom polythiophenes. *Nanoscale* 3: 4261–4269.

53. Horowitz, G. 1998. Organic field-effect transistors. *Adv. Mater.* 10: 365.

54. Nam, S., Ko, Y. G., Hahm, S. G., Park, S., Seo, J., Lee, H., Kim, H., Ree, M., and Kim, Y. 2013. Organic nonvolatile memory transistors with self-doped polymer energy well structures. *NPG Asia Mater.* 5: e33.

55. Han, H., Nam, S., Seo, J., Lee, C., Kim, H., Bradley, D. D. C., Ha, C.-S., and Kim, Y. 2015. Broadband all-polymer phototransistors with nanostructured bulk heterojunction layers of NIR-sensing n-type and visible light-sensing p-type polymers. *Sci. Rep.* 5: 16457.

56. Seo, J., Park, S., Nam, S., Kim, H., and Kim, Y. 2013. Liquid crystal-on-organic field-effect transistor sensory devices for perceptive sensing of ultralow intensity gas flow touch. *Sci. Rep.* 3: 2452.

57. Seo, J., Nam, S., Jeong, J., Lee, C., Kim, H., and Kim, Y. 2015. Liquid crystal-gated-organic field-effect transistors with in-plane drain–source–gate electrode structure. *ACS Appl. Mater. Interfaces* 7: 504–510.

58. Seo, J., Lee, C., Han, H., Lee, S., Nam, S., Kim, H., Lee, J. H., Park, S. Y., Kang, I. K., and Kim, Y. 2014. Touch sensors based on planar liquid crystal-gated-organic field-effect transistors. *AIP Adv.* 4: 097109.
59. Seo, J., Song, M., Han, H., Kim, H., Lee, J. H., Park, S. Y., Kang, I. K., and Kim, Y. 2015. Ultrasensitive tactile sensors based on planar liquid crystal-gated-organic field-effect transistors with polymeric dipole control layers. *RSC Adv.* 5: 56904.
60. Seo, J., Song, M., Lee, C., Nam, S., Kim, H., Park, S. Y., Kang, I. K., Lee, J. H., and Kim, Y. 2016. Physical force-sensitive touch responses in liquid crystal-gated-organic field-effect transistors with polymer dipole control layers. *Org. Electron.* 28: 184–188.

7 Liquid Crystals in Microfluidic Devices for Sensing Applications

Kun-Lin Yang

CONTENTS

7.1 Introduction 145
7.2 Microfluidic Devices 146
7.3 Detection of Surface-Bound Molecules with an LC 147
7.4 Orientations of an LC in Microchannels 149
7.5 LC in Microchannels for Sensing Applications 150
7.6 Advantages of Microfluidic LC Sensors 151
7.7 Cholesteric LC for Gas Detection 153
7.8 LC with Tubing Cartridges 154
7.9 Conclusion 155
References 155

7.1 INTRODUCTION

Microfluidic sensors possess several advantages such as smaller sample volume, shorter assay time, and lower cost compared with conventional assays performed in 96-well plates. However, because of its miniaturized format, microfluidic sensors also face serious challenges in signal detection because of the inherent short light path length, small sample volume, and short residence time. To overcome these limitations in microfluidic sensors, ultrasensitive signal detection and amplification mechanism are often required. In the past, laser-induced fluorescence (LIF) detection and enzymatic signal enhancement were the preferred detection methods in microfluidic sensors because of their high sensitivity. In these methods, target molecules need to be conjugated with labels such as fluorophores (Bernard et al. 2001; Hosokawa et al. 2007), enzymes (Eteshola and Balberg 2004; Yu et al. 2009), nanoparticles (Lu et al. 2009; Luo et al. 2005), and redox labels (Lim et al. 2003) to transduce detection signals. Nevertheless, labeling target molecules is tedious and the functionality of the molecules can be compromised upon conjugation with labels (Shrestha et al. 2012; Thorek et al. 2009). More importantly, if bulky instruments such as laser and fluorescence detectors are needed, then it becomes very difficult to miniaturize these microfluidic sensor devices. A better sensing mechanism fully compatible with a microfluidic device is therefore needed.

Liquid crystal (LC) has been used as a signal transducer for label-free detection with extremely high sensitivity (Gupta et al. 1998). It has been reported that long-range orientation of an LC responds sensitively to surface molecular binding events and changes the optical appearance of an LC, that is, a bright signal upon polar orientation shift (Alino and Yang 2011; Chen and Yang 2012) or change of the brightness upon azimuthal orientation shift (Govindaraju et al. 2007; Kim et al. 2000). These changes enable the development of label-free chemical and biological sensors for naked-eye detection. It is ideal to use an LC as a sensing material in microfluidic sensors because of following reasons. (1) The optical signal from an LC is visible to the naked eye under ambient light. Thus, bulky laser and emission filters are not needed. Moreover, an LC exhibits different colors which can be used to quantitatively determine concentrations of analytes. (2) An LC has good spatial resolution over a large area such that one can probe different locations on the microfluidic device with an LC. (3) An LC is highly sensitive as it can amplify molecular binding events into ordering transition (Gupta and Abbott 1997; Gupta et al. 1998; Shah and Abbott 2001). This feature is necessary for the ultrasensitive detection of molecules in a microfluidic device. (4) An LC is fluidic in nature and it can freely flow inside microchannels. Because it is hydrophobic, it can also displace water from the microchannel. Thanks to these unique features, an LC shows promise as an ideal sensing material for microfluidic devices. In this chapter, recent advances and efforts in this area are reviewed and discussed.

7.2 MICROFLUIDIC DEVICES

Microfluidic devices are characterized by their small dimensions, usually in the range of 1–1000 µm. Because of their small dimension, both mass transfer and heat transfer can be very fast. Thus, they are suitable for heterogeneous bioassays and biosensors in which binding of target analytes to a surface sensitive layer is required. These small devices are often made of transparent polymers such as polydimethylsiloxane (PDMS) or polymethylmethacrylate (PMMA), which allow optical signals to be transmitted through the microfluidic device and recorded. Applications of microfluidic devices for medical diagnosis have been demonstrated in many papers (Chin et al. 2011; Linder et al. 2005; Ikami et al. 2010). However, a common element in these devices is that the final detection step is still accomplished by using traditional methods, such as UV/Vis, fluorescence or electrical detection. During the early development stages, the detection was done by connecting a microfluidic device to mass spectrometers (Oleschuk and Harrison 2000; Koster and Verpoorte 2007), UV/Vis absorbance spectrometer, or LIF fluorometers (Gotz and Karst 2007). However, detection using these analytical instruments has limited portability. Another issue is that the path length of a microchannel is only 10–500 µm. Therefore, absorbance or fluorescence emission is much weaker than in conventional assays (according to Beer–Lambert law, absorbance is proportional to the optical path length). Thus, ancillary devices such as powerful light sources and ultrasensitive detectors are used. However, these bulky items easily compromise the advantages of microfluidic devices, since these ancillary items can be several times bigger than the microfluidic device. Alternatively, microelectrodes can be implanted inside the microfluidic devices as a sensor to report analyte concentration. However, it is difficult to fabricate these microelectrodes inside microchannels, and only a single reading can be obtained from an electrode. It is also difficult to obtain a concentration profile inside a microchannel by using electrical methods. Apparently, a new detection mechanism is critically needed for the microfluidic devices. A promising solution is to use an LC in microfluidic devices.

7.3 DETECTION OF SURFACE-BOUND MOLECULES WITH AN LC

When a very thin layer of an LC is supported on a solid surface, orientations of the LC are determined by several physical and chemical properties of the surface. This phenomenon is called "anchoring" of the LC. The average orientation of an LC near the surface is known as the easy axis of the surface (Figure 7.1). Factors including nanostructures, surface functional groups, and wettability are known to affect the easy axis. In the literature, various strategies have been used to control the anchoring of an LC and the easy axis of the surface. For example, oblique deposition of gold on a glass slide was used to create a planar easy axis which is parallel to the surface. The direction of the easy axis can be further controlled by using various self-assembled monolayers (SAMs) of alkanethiols on the gold surface (Gupta and Abbott 1996a,b, 1997). One can also control the easy axis of the surface by coating the glass slides with organosilanes such as N, N-dimethyl-n-octadecyl-3-aminopropyltrimethoxysilyl chloride (DMOAP), which gives a strong homeotropic anchoring of an LC (Kahn 1973). Alternatively, mechanical rubbing (Kim and Abbott 2001; Kim et al. 2000), rolling (Tingey et al. 2004), or surface periodic structures can be used to obtain planar anchoring (Chiou, Chen, and Lee 2006; Chiou, Yeh, and Chen 2006).

Detection of surface-bound molecules by using an LC often relies on changing of the easy axis of an LC after the binding of target analytes. This technique was first developed by Abbott's group in the late 1990s and gradually evolved into different formats over the past 20 years. In the earlier format, obliquely deposited

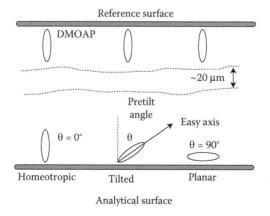

FIGURE 7.1 An LC optical cell with an analytical surface and a reference surface.

gold surfaces decorated with SAMs were used. These surfaces tend to align an LC uniformly with a single easy axis in one direction due to its unique surface topography and surface functionality (Luk, Yang et al. 2004). However, after the binding of molecules to the surfaces, the surface-bound molecules mask the surface topography and change the physical properties of the surface. As a result, the surfaces can no longer align the LC uniformly which led to random textures. For example, Gupta et al. (1998) showed that binding of proteins to ligands on an obliquely deposited gold surface changed the easy axis of the surface and the anchoring of the LC near the surface. Due to long-range interactions of the LC, this binding event then triggered changes in the orientation of a 20-μm LC film supported on the surface, giving distinct optical outputs. This technique was further developed for the detection of other proteins (Luk, Abbott, Bertics et al. 2003, Luk, Abbott, Raines et al. 2003; Luk, Tingey et al. 2003, 2004) and other applications such as the detection of acid–base reactions (Shah and Abbott 1999) or chemical vapors (Shah and Abbott 2003; Yang et al. 2004, 2005).

Alternatively, surface-bound molecules can also cause a homeotropic easy axis of a surface to become tilted or planar. In this case, a DMOAP-coated slide is commonly used as a substrate because it has a homeotropic easy axis with a strong anchoring energy. The DMAOP surface can also be modified with plasma or UV light and introduce reactive functional groups to immobilize proteins (Xue and Yang 2007, 2010). Xue et al. (2009) showed that adsorption of proteins on a DMOAP-coated slide can alter the homeotropic easy axis when the protein density exceeded a critical value. Chen and Yang (2010) showed that adsorption of DNA on the DMOAP surface also led to change in the easy axis.

However, characterizing the easy axis of a surface before and after surface binding events by using an LC is nontrivial. This is because an LC flows like a liquid on a surface, and it does not form a flat film due to surface tension. To observe the easy axis of a surface and the orientation of an LC, one has to pair an analytical surface with a reference surface (with a known easy axis) to form an optical cell (for filling with an LC) as shown in Figure 7.1. In many studies, a DMOAP-coated slide was selected as a reference surface because the easy axis is perpendicular to the surface which leads to the homeotropic anchoring of LC. In other studies, an obliquely deposited gold surface with a uniform planar easy axis was selected as a reference surface (Lowe et al. 2008). The two surfaces (analytical and reference) can be separated by a pair of spacer which defines the thickness of the LC film.

To probe the easy axis of the analytical surface, a polarized optical microscope (POM) is needed to determine the effective birefringence of the LC film. By comparing the interference color of the LC with the Michel-Levy chart, the effective birefringence of the LC can be readily obtained. The effective birefringence reflects the director profile of the LC in the entire film. If both easy axes of the analytical and reference surfaces are homeotropic, then the director profile of the LC will be uniformly homeotropic throughout the film. Under crossed polars, the sample always appears dark even when the sample is under rotation. This uniform homeotropic state can be further confirmed as a dark cross under a conoscopic lens. In this case, if the easy axis on the reference surface is homeotropic, then one can conclude that the easy axis of the analytical surface is also homeotropic (unless the cell thickness is less than a critical value, making the reference surface dominate the orientation of the LC).

In contrast, if the LC film is bright and shows modulation of light under crossed polars, then the easy axis on the analytical surface must be planar or tilted with a pretilt angle θ (the angle between surface norm and the director of an LC). In this case, the director profile of the LC will undergo some bend and splay distortion in the film. A mathematical model proposed by Barbero and Barberi (1983) was used to predict the director profile of an LC in such hybrid cells. From this model, the measured effective birefringence can be used to determine the director profile and pretilt angle θ on the surface, which reveals the actual easy axis on the analytical surface. One can compare the easy axis before and after the binding events to decide if the binding of molecules changes the easy axis.

It is worth noting that in the technique mentioned above, the thickness of the LC film is around 1–20 mm, which is comparable with the dimension of a typical microfluidic device. Moreover, the signal is not proportional to the thickness of the LC film since changes of easy axis and anchoring of LC are surface-driven phenomena.

7.4 ORIENTATIONS OF AN LC IN MICROCHANNELS

When an LC is inside a rectangular microchannel for sensing applications, it is surrounded by four channel walls and its director profile is controlled by the anchoring energies of all surfaces and the elastic energy in an LC. However, in a rectangular channel with a large aspect ratio ($w/d > 10$), one can assume that the orientational profile of an LC is dictated by the top and bottom surface only. Thus, all techniques developed for LC optical cells are also applicable for an LC in microchannels. For example, when both the top and bottom surfaces have homeotropic conditions, the orientation profile of an LC in the microchannel is also uniformly homeotropic, which can be determined by using a POM coupled with conoscopic lens.

In the case of square microchannels or rectangular microchannels with a small aspect ratio (w/d), the orientational profile of LC is more complicated since LC is surrounded by four channel walls. In a square microchannel with four homeotropic boundary conditions, the orientation of the LC in the middle of the microchannel is parallel to the length of the microchannel rather than being perpendicular to any of the wall. To obtain the actual orientational profile of an LC in these microchannels, the LC can be doped with a fluorescent dye molecule and examined by using a POM and a fluorescent confocal polarized microscopy (FCPM) as shown in a study by Sengupta et al. (2013). The orientational profile of an LC in a microchannel can also be predicted by using numerical simulation.

Sengupta et al. used both experiments and computer simulations to investigate the orientational profile of an LC 5CB in a rectangular microchannel ($w \times d = 100\,\mu m \times 10\,\mu m$) with homeotropic boundary conditions. Due to a large aseptic ratio, 5CB in the middle of the cell aligns perpendicularly to the top and bottom surfaces, whereas 5CB at both sides aligns horizontally with various bending and splaying to satisfy the boundary conditions on the side (Zhu and Yang 2015).

When an LC is flowing in a microfluidic channel, the orientational profile of the LC is strongly influenced by the flow as suggested by Sengupta et al. (2013). Under a weak-flow regime, the orientational profile of an LC is similar to that of the no-flow regime since the flow has minimal effect on the LC. However, when the flow speed of the LC increases and enters a medium-flow regime, the orientational profile of the LC is influenced by the flow and vice versa. A complex rheology caused by competition between the elasticity and the backflow can be observed. For example, the orientational profile becomes symmetrical along the center of the microchannel, and the flow changes to a two-stream flow with a lower velocity at the channel center. Eventually, at the strong-flow regime, the flow dominates the orientations of the LC, making the director the same as the flowing direction. Such a phenomenon is unique to anisotropic LCs and may have important implications for sensing applications in the future.

In a recent study, spherical 5CB droplets flowing inside microchannels filled with water was investigated. In this case, the orientations of 5CB are determined by water rather than channel wall surface. Since water imposes a planar boundary condition on the 5CB droplets, the orientational profile of 5CB can be bipolar, concentric, or escaped concentric. In most cases, only a bipolar configuration is observed due to its lower energy state. Nevertheless, when a 5CB droplet started to flow inside the channel, backflow of 5CB affected the location of defects such that the droplet changed cyclically between bipolar and escaped concentric configurations. Since the orientational profile of 5CB is very sensitive to the flow velocity and shear stress, it can be used to report both parameters in a microfluidic channel.

7.5   LC IN MICROCHANNELS FOR SENSING APPLICATIONS

To combine an LC with microfluidic devices for sensing applications, several formats have been used in the past. The simplest format is to perform biological assays in a traditional PDMS microfluidic device (supported on a glass slide) and then disassemble the device to expose the bottom glass slide. To detect whether the molecules bind to the bottom surface by using LC, a DMOAP-coated slide can be paired with the bottom surface as an optical cell to fill the LC. Xue et al. (2009) used this approach to build an LC-based microfluidic immunoassay for detecting various antibodies with high specificity. However, it is challenging to disassemble the microfluidic device without affecting the bottom surface. The bottom surface also needs to be rinsed to avoid nonspecific adsorption of proteins on the surface (Zhang et al. 2011). A more straightforward method is to incorporate the LC inside microchannels as a sensor to detect target analytes directly. However, an LC is fluid in nature and unstable under flow conditions. To use an LC in microchannels over an extended period of time, the LC has to be properly confined inside the microchannels.

A popular method to confine LC is by using a transmission electron microscopy (TEM) grid. Due to strong capillary force, an LC can be confined in the TEM grid. This technique was developed by Brake and Abbott (2002) to study the adsorption of surfactants or biomolecular binding events at LC/water interfaces. This technique was used in other studies for the development of chemical and biological sensors (Hartono et al. 2008; Hartono, Qin et al. 2009; Hartono, Xue et al. 2009). However, it is difficult to dispense the exact amount of LC into the grid (since the grid is very small). In some cases, excess LC needs to be removed manually by touching the

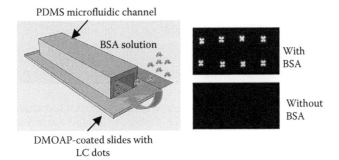

FIGURE 7.2 Inkjet-printed LC sensing dots in microchannels for the detection of bovine serum albumin (BSA). (Reprinted with permission from Alino, V. J. et al. 2012. Inkjet printing and release of monodisperse liquid crystal droplets from solid surfaces. *Langmuir* 28 (41):14540–14546. Copyright 2012 American Chemical Society.)

TEM grid with a glass capillary. Moreover, it is not feasible to incorporate TEM grids into a microfluidic device, because the size of a TEM grid is larger than the dimension of a microchannel. To address these issues mentioned above, Liu et al. (2012) demonstrated a highly reproducible technique to form a stable LC film inside a microchannel as follows. First, a hexagonal grid was electroplated inside a micro-channel using a photomask. Next, the LC was infused into the channel to form a stable film inside the grid. An advantage of this technique is that excess LC can be removed by a laminar flow of water without any manual steps. This technique led to a highly reproducible, long-lasting LC-sensing film in the microchannel, which can be used for various sensing applications such as detection of surfactants or enzyme phospholipase A_2.

Alternatively, tiny LC droplets can be printed inside a microchannel for sensing applications. For instance, Alino et al. (2012) showed that LC 5CB can be printed by using a piezoelectric nozzle at an elevated temperature (to lower its viscosity). Subsequently, LC droplets printed on a solid surface were covered with microchannels and formed tiny LC sensing dots in the microchannels. Depending on surface chemistry and flow rate of the solution, the LC sensing dots can be very stable for sensing purposes. By using these techniques, Alino et al. demonstrated that detection of surfactants and proteins can be accomplished by using a microfluidic device and LC sensing dots as shown in Figure 7.2. However, these sensing dots may detach from the surface if the flow rate is too high.

7.6   ADVANTAGES OF MICROFLUIDIC LC SENSORS

An important consideration for developing a useful immunoassay is its ability to quantify antibody concentrations. In microfluidic immunoassays, the quantification part can be realized by using absorbance of light or the intensity of fluorescence. However, because the dimension of microchannels is very small, and the absorbance is proportional to the path length of light, sensitivity of the microfluidic sensor is low. On the other hand, the detection principle of an LC is not limited by the small

FIGURE 7.3 An LC-based immunoassay developed in microfluidic channels. The surface of the slide was coated with IgG, and a buffer containing anti-IgG was infused into the first channel. The bright region is where anti-IgG in buffer solution binds to surface-bound IgG. The amount of anti-IgG that binds to the surface can be estimated by using the brightness of the LC. (Reprinted with permission from Xue, C.-Y., S. A. Khan, and K.-L. Yang. 2009. *Advanced Materials* 21 (2):198–202.)

dimensions of microchannels. Since orientations of LC are disrupted by surface-bound molecules to produce interference colors, the sensitivity of LC-based detection actually increases with decreasing thickness of the LC film. The concept of using an LC to detect binding of anti-IgG to surface-bound IgG quantitatively in a microfluidic device has been reported by Xue et al. In this study, binding of anti-IgG to IgG was evident from the interference colors of LC as shown in Figure 7.3. After the binding of anti-IgG to surface-bound IgG, the brightness of the LC followed a decreasing order: from white, gray and then to black along the channel. This phenomenon was caused by a gradual decrease in the pretilt angle of the LC along the microchannel. This is because anti-IgG is gradually depleted inside the microchannel when it binds to immobilized IgG on the surface. Therefore, the surface density of anti-IgG in the inlet is higher than the other regions. The high density of anti-IgG led to a large pretilt angle of the LC. Based on this principle, one can predict the antibody concentration in each region by using the brightness or pretilt angles of the LC. This is one of the advantages of using an LC for detection, since other detection methods (e.g., fluorescence or nanoparticles) do not have such a feature. Nevertheless, the pretilt angle of the LC is only suitable for quantitative analysis, and it is very sensitive to the thickness of the LC. A small difference in the cell thickness or force imbalance can lead to a bias.

Alternatively, quantification mechanism can be achieved by using the length of bright LC region in the microfluidic channels. Because the LC is highly sensitive to the critical concentration (Xue and Yang 2008), a clear cutoff can be found at the end of the bright channel as shown Figure 7.4. Therefore, the length of the bright LC region can be measured more precisely, and the length is proportional to the antibody concentration. In contrast, fluorescence intensity decreases continuously, making it impossible to determine the length of fluorescent region. This technique has been commonly used in LC-based assays for quantification purposes (Chen and Yang 2012; Zhang et al. 2011; Zhu and Yang 2015).

Another advantage of using an LC in microfluidic devices lies in its spatial resolution. Users can probe the entire surface with an LC to find out the distribution of analytes on the surface. Because of this feature, it is feasible to use an LC for multiplex and high-throughput detection in microfluidic devices. An LC image of a microfluidic immunoassay in Figure 7.5 shows that only line–line intersections

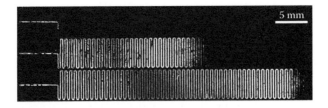

FIGURE 7.4 (**See color insert.**) Correlations between the length of the bright LC regions and the concentration of anti-IgG. From the top, the concentration of anti-IgG was 0.02, 0.05, and 0.08 mg/mL, respectively.

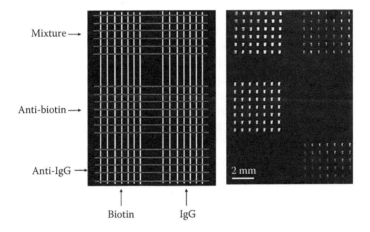

FIGURE 7.5 A high-throughput and multiplex LC immunoassay. The LC only appears bright in the line–line intersections where antigens and their respective antibody meet. In the case of mixtures which contain both anti-biotin and anti-IgG, LCs in both biotin-BSA and IgG channels appear bright.

where antibodies meet their specific target proteins light up, which suggests that the specific binding of both antigen/antibody pairs can be detected with LCs in a multiplexed manner. The brightness of the LC can also be used to determine the binding strength of the antigen/antibody pair. As shown in Figure 7.5, since the binding of antibiotin to biotin-BSA is stronger than the binding of anti-IgG to IgG, the brightness of the former is also more pronounced.

7.7 CHOLESTERIC LC FOR GAS DETECTION

In addition to nematic LC, cholesteric liquid crystals (CLCs) have also been used in microfluidic devices for gas sensing applications. CLCs are a type of LCs whose molecules are arranged in layers. The orientation of each layer rotates a small angle to form a helical pattern. CLCs selectively reflect light according to Bragg's law when the pitch matches the incident light. Thanks to this property, CLCs are very sensitive to temperature and volatile organic compounds (VOCs) (Dickert et al. 1992; Dickert et al. 1994; Sutarlie et al. 2011; Sutarlie et al. 2010). For example, when CLCs are exposed to VOCs, dissolution of VOCs changes the pitch of CLCs and results in different colors. Unlike nematic LCs, the colors of CLCs can be viewed with the naked eye directly without the use of polarizers. However, CLCs are not specific to VOCs. To make them selective to target molecules, reactive dopants were often mixed with CLCs to render specificity. For example, dodecylamine was mixed with CLCs to detect 300 ppmv of pentyl aldehyde vapor since the primary amine is able to react with the aldehyde and form imines (Sutarlie et al. 2011). Because CLCs are very viscous, they can be patterned on a surface directly as sensing arrays for detecting different types of VOCs.

CLCs also have a good spatial resolution when they are used as sensing materials, making them very suitable to be integrated with microfluidic devices for sensing applications. This can be done by applying a thin layer of CLCs directly to microfluidic devices and then protecting it with a layer of gas-permeable membrane. In this case, gas molecules can permeate through the membrane and interact with CLCs to change their colors. Alternatively, CLCs can be blended with polymers to make a polymer dispersed cholesteric liquid crystal (PDCLC). A film of PDCLC can be used to cover a microfluidic device for sensing VOCs (Sutarlie and Yang 2011). This method was used to detect the distribution of 4% ethanol solution inside microfluidic channels. Ethanol produced from fermentation in microfluidic reactors can also be detected.

7.8 LC WITH TUBING CARTRIDGES

In most LC-based immunoassays, manual preparation of an LC cell for detection is required. This is because a number of incubation and rinsing steps are needed before the test results can be obtained. Unfortunately, aqueous buffers are not miscible with LCs. Therefore, it is more convenient to introduce an LC at the final step after all the incubation and rinsing steps are completed. Recently, a new format of LC-based microfluidic immunoassay was reported by Zhu (Zhu and Yang 2015). In this assay, an LC was injected directly into a microchannel at the end of the immunoassay procedures. Because LC and water are immiscible, the LC is able to push out all remaining solutions from the microchannel and probe the analytical surface. Depending on the binding of the antibody to the surface, the LC may appear bright or dark. However, orientations of the LC are influenced by four channel walls. Thus, the dimension of the microchannel needs to be properly controlled.

To streamline the process mentioned above, Zhu and Yang prepared a tubing cartridge which contains "plugs" of antibody solution, a washing buffer, and an LC. These plugs are separated by an air spacer, ensuring that they are not mixed together. This setup was first invented by the Whitesides' group, but no LC was used (Chin et al. 2011; Linder et al. 2005). To drive the solutions and LC into the channel, the cartridge was connected to the inlet of the microchannel, and negative pressure (a syringe barrel) was connected to the outlet to create a pressure-driven flow. The flow velocity can be controlled very accurately by varying the position of the plunger. Volume of the antibody solution and washing buffer can also be preprogrammed to achieve optimum performance.

7.9 CONCLUSION

LCs have been integrated with microfluidic devices for various chemical and biological sensing applications. Using LCs for signal transduction and amplification is advantageous because it is highly sensitive, fast, and the optical outputs can be observed directly with the naked eye. Interference colors of LCs also offer a simple way for qualitative and spatial analysis of protein concentration. Despite their tremendous advantages, the best platform to use LCs in microfluidic devices for sensing applications is still under debate. In the future, more research should be carried out in this direction to optimize LC-based microfluidic sensing devices for real-world applications.

REFERENCES

Alino, V. J., K. X. Tay, S. A. Khan, and K.-L. Yang. 2012. Inkjet printing and release of monodisperse liquid crystal droplets from solid surfaces. *Langmuir* 28 (41):14540–14546.

Alino, V. J. and K.-L. Yang. 2011. Using liquid crystals as a readout system in urinary albumin assays. *Analyst* 136 (16):3307–3313.

Barbero, G. and R. Barberi. 1983. Critical thickness of a hybrid aligned nematic liquid-crystal cell. *Journal De Physique* 44 (5):609–616.

Bernard, A., B. Michel, and E. Delamarche. 2001. Micromosaic immunoassays. *Analytical Chemistry* 73 (1):8–12.

Brake, J. M. and N. L. Abbott. 2002. An experimental system for imaging the reversible adsorption of amphiphiles at aqueous-liquid crystal interfaces. *Langmuir* 18 (16):6101–6109.

Chen, C.-H. and K.-L. Yang. 2010. Detection and quantification of DNA adsorbed on solid surfaces by using liquid crystals. *Langmuir* 26 (3):1427–1430.

Chen, C.-H. and K.-L. Yang. 2012. Liquid crystal-based immunoassays for detecting hepatitis B antibody. *Analytical Biochemistry* 421 (1):321–323.

Chin, C. D., T. Laksanasopin, Y. K. Cheung, D. Steinmiller, V. Linder, H. Parsa, J. Wang et al. 2011. Microfluidics-based diagnostics of infectious diseases in the developing world. *Nature Medicine* 17 (8):1015–1019.

Chiou, D. R., L. J. Chen, and C. D. Lee. 2006. Pretilt angle of liquid crystals and liquid-crystal alignment on microgrooved polyimide surfaces fabricated by soft embossing method. *Langmuir* 22 (22):9403–9408.

Chiou, D. R., K. Y. Yeh, and L. J. Chen. 2006. Adjustable pretilt angle of nematic 4-n-pentyl-4′-cyanobiphenyl on self-assembled monolayers formed from organosilanes on square-wave grating silica surfaces. *Applied Physics Letters* 88 (13).

Dickert, F. L., A. Haunschild, and P. Hofmann. 1994. Cholesteric liquid-crystals for solvent vapor detection—elimination of cross-sensitivity by band shape-analysis and pattern-recognition. *Fresenius Journal of Analytical Chemistry* 350 (10–11):577–581.

Dickert, F. L., A. Haunschild, P. Hofmann, and G. Mages. 1992. Molecular recognition of organic-solvents and ammonia: Shapes and donor properties as sensor effects. *Sensors and Actuators B-Chemical* 6 (1–3):25–28.

Eteshola, E. and M. Balberg. 2004. Microfluidic ELISA: On-chip fluorescence imaging. *Biomedical Microdevices* 6 (1):7–9.

Gotz, S. and U. Karst. 2007. Recent developments in optical detection methods for microchip separations. *Analytical and Bioanalytical Chemistry* 387 (1):183–192.

Govindaraju, T., P. J. Bertics, R. T. Raines, and N. L. Abbott. 2007. Using measurements of anchoring energies of liquid crystals on surfaces to quantify proteins captured by immobilized ligands. *Journal of the American Chemical Society* 129 (36):11223–11231.

Gupta, V. K. and N. L. Abbott. 1996a. Azimuthal anchoring transition of nematic liquid crystals on self-assembled monolayers formed from odd and even alkanethiols. *Physical Review E* 54 (5):R4540–R4543.

Gupta, V. K. and N. L. Abbott. 1996b. Uniform anchoring of nematic liquid crystals on self-assembled monolayers formed from alkanethiols on obliquely deposited films of gold. *Langmuir* 12 (10):2587–2593.

Gupta, V. K. and N. L. Abbott. 1997. Design of surfaces for patterned alignment of liquid crystals on planar and curved substrates. *Science* 276 (5318):1533–1536.

Gupta, V. K., J. J. Skaife, T. B. Dubrovsky, and N. L. Abbott. 1998. Optical amplification of ligand-receptor binding using liquid crystals. *Science* 279 (5359):2077–2080.

Hartono, D., X. Bi, K.-L. Yang, and L.-Y. L. Yung. 2008. An air-supported liquid crystal system for real-time and label-free characterization of phospholipases and their inhibitors. *Advanced Functional Materials* 18 (19):2938–2945.

Hartono, D., W. J. Qin, K.-L. Yang, and L.-Y. L. Yung. 2009. Imaging the disruption of phospholipid monolayer by protein-coated nanoparticles using ordering transitions of liquid crystals. *Biomaterials* 30 (5):843–849.

Hartono, D., C.-Y. Xue, K.-L. Yang, and L.-Y. L. Yung. 2009. Decorating liquid crystal surfaces with proteins for real-time detection of specific protein-protein binding. *Advanced Functional Materials* 19 (22):3574–3579.

Hosokawa, K., M. Omata, and M. Maeda. 2007. Immunoassay on a power-free microchip with laminar flow-assisted dendritic amplification. *Analytical Chemistry* 79 (15):6000–6004.

Ikami, M., A. Kawakami, M. Kakuta, Y. Okamoto, N. Kaji, M. Tokeshi, and Y. Baba. 2010. Immuno-pillar chip: A new platform for rapid and easy-to-use immunoassay. *Lab on a Chip* 10 (24):3335–3340.

Kahn, F. J. 1973. Orientation of liquid-crystals by surface coupling agents. *Applied Physics Letters* 22 (8):386–388.

Kim, S. R. and N. L. Abbott. 2001. Rubbed films of functionalized bovine serum albumin as substrates for the imaging of protein–receptor interactions using liquid crystals. *Advanced Materials* 13 (19):1445–1449.

Kim, S. R., R. R. Shah, and N. L. Abbott. 2000. Orientations of liquid crystals on mechanically rubbed films of bovine serum albumin: A possible substrate for biomolecular assays based on liquid crystals. *Analytical Chemistry* 72 (19):4646–4653.

Koster, S. and E. Verpoorte. 2007. A decade of microfluidic analysis coupled with electrospray mass spectrometry: An overview. *Lab on a Chip* 7 (11):1394–1412.

Lim, T. K., H. Ohta, and T. Matsunaga. 2003. Microfabricated on-chip-type electrochemical flow immunoassay system for the detection of histamine released in whole blood samples. *Analytical Chemistry* 75 (14):3316–3321.

Linder, V., S. K. Sia, and G. M. Whitesides. 2005. Reagent-loaded cartridges for valveless and automated fluid delivery in microfluidic devices. *Analytical Chemistry* 77 (1):64–71.

Liu, Y., D. M. Cheng, I. H. Lin, N. L. Abbott, and H. R. Jiang. 2012. Microfluidic sensing devices employing *in situ*-formed liquid crystal thin film for detection of biochemical interactions. *Lab on a Chip* 12 (19):3746–3753.

Lowe, A. M., P. J. Bertics, and N. L. Abbott. 2008. Quantitative methods based on twisted nematic liquid crystals for mapping surfaces patterned with bio/chemical functionality relevant to bioanalytical assays. *Analytical Chemistry* 80 (8):2637–2645.

Lu, Y., W. W. Shi, J. H. Qin, and B. C. Lin. 2009. Low cost, portable detection of gold nanoparticle-labeled microfluidic immunoassay with camera cell phone. *Electrophoresis* 30 (4):579–582.

Luk, Y. Y., N. L. Abbott, P. J. Bertics, M. L. Tingey, and D. J. Hall. 2003. Using liquid crystals and nanostructured surfaces to detect regulatory proteins involved in cell signaling pathways. *Biochemistry* 42 (28):8638–8638.

Luk, Y. Y., N. L. Abbott, R. T. Raines, M. L. Tingey, and K. A. Dickson. 2003. Comparison of the binding activity of randomly oriented and uniformly oriented proteins immobilized by chemoselective coupling to a self-assembled monolayer. *Biochemistry* 42 (28):8650–8650.

Luk, Y. Y., M. L. Tingey, K. A. Dickson, R. T. Raines, and N. L. Abbott. 2004. Imaging the binding ability of proteins immobilized on surfaces with different orientations by using liquid crystals. *Journal of the American Chemical Society* 126 (29):9024–9032.

Luk, Y. Y., M. L. Tingey, D. J. Hall, B. A. Israel, C. J. Murphy, P. J. Bertics, and N. L. Abbott. 2003. Using liquid crystals to amplify protein–receptor interactions: Design of surfaces with nanometer-scale topography that present histidine-tagged protein receptors. *Langmuir* 19 (5):1671–1680.

Luk, Y. Y., K.-L. Yang, K. Cadwell, and N. L. Abbott. 2004. Deciphering the interactions between liquid crystals and chemically functionalized surfaces: Role of hydrogen bonding on orientations of liquid crystals. *Surface Science* 570 (1–2):43–56.

Luo, C. X., Q. Fu, H. Li, L. P. Xu, M. H. Sun, Q. Ouyang, Y. Chen, and H. Ji. 2005. PDMS microfluidic device for optical detection of protein immunoassay using gold nanoparticles. *Lab on a Chip* 5 (7):726–729.

Oleschuk, R. D. and D. J. Harrison. 2000. Analytical microdevices for mass spectrometry. *Trac-Trends in Analytical Chemistry* 19 (6):379–388.

Sengupta, A., U. Tkalec, M. Ravnik, J. M. Yeomans, C. Bahr, and S. Herminghaus. 2013. Liquid crystal microfluidics for tunable flow shaping. *Physical Review Letters* 110 (4).

Shah, R. R. and N. L. Abbott. 1999. Using liquid crystals to image reactants and products of acid-base reactions on surfaces with micrometer resolution. *Journal of the American Chemical Society* 121 (49):11300–11310.

Shah, R. R. and N. L. Abbott. 2001. Principles for measurement of chemical exposure based on recognition-driven anchoring transitions in liquid crystals. *Science* 293 (5533):1296–1299.

Shah, R. R. and N. L. Abbott. 2003. Orientational transitions of liquid crystals driven by binding of organoamines to carboxylic acids presented at surfaces with nanometer-scale topography. *Langmuir* 19 (2):275–284.

Shrestha, D., A. Bagosi, J. Szollosi, and A. Jenei. 2012. Comparative study of the three different fluorophore antibody conjugation strategies. *Analytical and Bioanalytical Chemistry* 404 (5):1449–1463.

Sutarlie, L., J. Y. Lim, and K.-L. Yang. 2011. Cholesteric liquid crystals doped with dodecylamine for detecting aldehyde vapors. *Analytical Chemistry* 83 (13):5253–5258.

Sutarlie, L., H. Qin, and K.-L. Yang. 2010. Polymer stabilized cholesteric liquid crystal arrays for detecting vaporous amines. *Analyst* 135 (7):1691–1696.

Sutarlie, L. and K.-L. Yang. 2011. Monitoring spatial distribution of ethanol in microfluidic channels by using a thin layer of cholesteric liquid crystal. *Lab on a Chip* 11 (23):4093–4098.

Thorek, D. L. J., D. R. Elias, and A. Tsourkas. 2009. Comparative analysis of nanoparticle-antibody conjugations: Carbodiimide versus click chemistry. *Molecular Imaging* 8 (4):221–229.

Tingey, M. L., E. J. Snodgrass, and N. L. Abbott. 2004. Patterned orientations of liquid crystals on affinity microcontact printed proteins. *Advanced Materials* 16 (15):1331–1336.

Xue, C.-Y., S. A. Khan, and K.-L. Yang. 2009. Exploring optical properties of liquid crystals for developing label-free and high-throughput microfluidic immunoassays. *Advanced Materials* 21 (2):198–202.

Xue, C.-Y. and K.-L. Yang. 2007. Chemical modifications of inert organic monolayers with oxygen plasma for biosensor applications. *Langmuir* 23 (10):5831–5835.

Xue, C.-Y. and K.-L. Yang. 2008. Dark-to-bright optical responses of liquid crystals supported on solid surfaces decorated with proteins. *Langmuir* 24 (2):563–567.

Xue, C.-Y. and K.-L. Yang. 2010. One-step UV lithography for activation of inert hydro-carbon monolayers and preparation of protein micropatterns. *Journal of Colloid and Interface Science* 344 (1):48–53.

Yang, K.-L., K. Cadwell, and N. L. Abbott. 2004. Mechanistic study of the anchoring behav-ior of liquid crystals supported on metal salts and their orientational responses to dimethyl methylphosphonate. *Journal of Physical Chemistry B* 108 (52):20180–20186.

Yang, K.-L., K. Cadwell, and N. L. Abbott. 2005. Use of self-assembled monolayers, metal ions and smectic liquid crystals to detect organophosphonates. *Sensors and Actuators B-Chemical* 104 (1):50–56.

Yu, L., C. M. Li, Y. S. Liu, J. Gao, W. Wang, and Y. Gan. 2009. Flow-through functional-ized PDMS microfluidic channels with dextran derivative for ELISAs. *Lab on a Chip* 9 (9):1243–1247.

Zhang, W., W. T. Ang, C.-Y. Xue, and K.-L. Yang. 2011. Minimizing nonspecific protein adsorption in liquid crystal immunoassays by using surfactants. *Acs Applied Materials & Interfaces* 3 (9):3496–3500.

Zhu, Q., and K.-L. Yang. 2015. Microfluidic immunoassay with plug-in liquid crystal for opti-cal detection of antibody. *Analytica Chimica Acta* 853:696–701.

Index

A

Alcohol sensors, 90–91; *see also* Optical sensors
All-electrical liquid crystal sensors, 103, 120; *see also* Liquid crystal; Nematic liquid crystal cell; Temperature–frequency converter
  electrical equivalent circuit, 108, 109–111
  experimental devices, 105
  LC capacitance temperature dependence, 111
  LC cell fabrication, 105–106
  permittivity of LC devices, 107–108
  sensing, 103
  temperature-phase converter, 116–120
Amine sensors, 86–88; *see also* Sensors
Amino acid sensors, 86; *see also* Sensors
Ancillary devices, 146; *see also* Microfluidic LC sensors
Astable multivibrator circuit, 111; *see also* Temperature–frequency converter
Azobenzene, 36
  derivatives, 35

B

Bandwidth tunable CLC, 1, 23; *see also* Broadband reflection; Cholesteric liquid crystals; Reflection bandwidth; Stimuli-responsive reflection band
  Bragg reflection wavelength, 1
  helical structure of CLC phase, 2
  mesogens, 1
BINOL (Bis-2-naphthol), 71
  forms of, 72
BINSH chiral dopant, 72
Blue phases (BPs), 35, 37
Bovine serum albumin (BSA), 151
BPs, *see* Blue phases
Bragg reflection wavelength, 1
Bragg's law, 83
Broadband reflection, 18; *see also* Bandwidth tunable CLC
  fractured cuticle of *Chrysina resplendens*, 20
  hyper-reflection of CLC, 23, 24
  mechanism of, 19
  multilayer solution, 20
  thermally induced handedness inversion of CLC, 21
  washout/refill techniques, 21, 22
BSA, *see* Bovine serum albumin

C

Carboxylate salt cholesteric films, 90
CD, *see* Circular dichroism
Chemosensors, 67; *see also* Chiral nematic liquid crystalline sensors
  BINOL forms, 72
  BINSH chiral dopant, 72
  chiral dopant and acetone reaction, 73
  crown-ether-functionalized cholesterol derivative, 68
  1,2-diphenylethane-1,2-diamine, 71
  humidity-sensitive BINOL dimer and H_2O, 73
  humidity sensor, 72
  hysteresis, 71
  nonchiral dopants, 73–74
  reactive cholesterol derivatives, 68–70
  reflection bands of cholesteric LCs, 74
  responsive chiral dopants, 70–73
  systems based on functionalized cholesterol derivatives, 70
  TFA-functionalized cholesterol derivatives, 69
Chiral azobenzene, 40, 41, 43, 44, 48; *see also* Photochromic chiral LCs
  light-driven, 51, 52
Chiral ionic liquid (CIL), 12
Chiral liquid crystalline phases, 36; *see also* Photochromic chiral LCs
Chiral nematic LC, 84; *see also* Cholesteric liquid crystalline polymer networks
Chiral nematic liquid crystalline sensors, 63, 79–80; *see also* Chemosensors; Sensors; Ultraviolet sensors
  mean refractive index, 64
  nematic mesophase, 64
  optical purity, 65
  pitch, 64, 66
  sensors, 65
Chiral nematic phase, 1
Chiral spirooxazine molecular switches, 45; *see also* Photochromic chiral LCs
Chiroptical switching, 76
Cholesteric, 63; *see also* Bandwidth tunable CLC
  diacrylate, 4
Cholesteric liquid crystalline polymer networks, 83, 97–98, 100; *see also* Optical sensors
  Bragg's law, 83
  chiral nematic LCl, 84

Cholesteric liquid crystalline polymer networks
    (*Continued*)
  hyper-reflection of, 23, 24
  pitch of chiral nematic, 83
  selective reflection band, 84
  thermally induced handedness inversion
    of, 21
Cholesteric liquid crystals (CLCs), 1, 23,
    34, 85, 153; *see also* Bandwidth
    tunable CLC
  light reflection property of, 2
  wide-band, 25
Cholesteric phase, 36
CIL, *see* Chiral ionic liquid
Circular dichroism (CD), 79
Circularly polarized light (CPL), 37
CLCs, *see* Cholesteric liquid crystals
Coating technologies, 125
Conventional sensors, 123
CPL, *see* Circularly polarized light

D

DBR, *see* Distributed Bragg reflection
DCL, *see* Dipole control layer
DCL-LC-*g*-OFET sensory devices, 135; *see also*
    LC-*i*-OFET sensory devices
  drain current and time, 136, 137
  drain current response, 139
  image comparison, optical
    microscope, 136
  I-V characteristics for diode
    devices, 135
  LC layer fluctuation in planar, 137
  PDMS bank and skin parts, 138
  physical touch load, 140
  structure of, 138
  transfer characteristics of, 138
  transfer curve for planar, 135
DFB, *see* Distributed feedback
DIAB (1,4-di-(4-(6-acryloxyhexyloxy)
    benzoyloxy) benzene), 85
Diarylethenes, 36
1,2-diphenylethane-1,2-diamine, 71
Dipole control layer (DCL), 135; *see also*
    DCL-LC-*g*-OFET sensory devices
Display devices, 34
Distributed Bragg reflection (DBR), 50
Distributed feedback (DFB), 48, 50
Dithienylcyclopentene, 36
DMOAP (*N*,*N*-dimethyl-*n*-octadecyl-3-
    aminopropyltrimethoxysilyl
    chloride), 147
Double-twist cylinders (DTCs), 37
DSCs, *see* Dye-sensitized solar cells
DTCs, *see* Double-twist cylinders
Dye-sensitized solar cells (DSCs), 25

E

Electrical equivalent circuit temperature
    dependence, 109–111
Enantiomeric light-driven dithienylethene chiral
    switches, 16

F

FCPM, *see* Fluorescent confocal polarized
    microscopy
Field-effect mobility, 126
Fluorescent confocal polarized microscopy
    (FCPM), 149

G

GNRs (Gold nanorods), 53

H

HBA ((6-hexaneoxy-4-benzoic acid) acrylate), 85
Helical twisting power (HTP), 2, 37, 65
Helicity inducers, 39
HTP, *see* Helical twisting power
Humidity sensor, 72, 89–90; *see also* Sensors

I

Indium-tin oxide (ITO), 105, 126
Induced N*LCs, 63
Infrared (IR), 3, 76
Internet-of things (IoT), 103
Interpenetrating network (IPN), 86
IoT, *see* Internet-of things
IPN, *see* Interpenetrating network
IR, *see* Infrared
ITO, *see* Indium-tin oxide

L

Laser-induced fluorescence (LIF), 145
LCD, *see* Liquid crystal display
LC-*g*-OFET sensory devices, 126, 129, 130; *see
    also* Organic field-effect transistors
  charge generation in P3HT layers, 135
  DCL-LC-*g*-OFET sensory devices, 135–140
  DCL to block charge generation, 134
  drain current change, 133
  hole mobility and threshold voltage, 131
  microscope images for, 132
  negative drain current, 133
  operation mechanism, 131
  structure of, 130, 133
  transfer curves, 130, 133
LC-*i*-OFET sensory devices, 124, 126; *see also*
    Organic field-effect transistors

LC-*on*-OFET sensory devices, 126–129; *see also* Organic field-effect transistors
 fabrication, 127
 gas flow-stimulated state in, 127
 output curves of OFET devices, 128
 sensing mechanism in, 128
 sensing performances of, 129
LCP, *see* Left-circularly polarized
Left-circularly polarized (LCP), 18
Left-handed (LH), 1
LH, *see* Left-handed
LIF, *see* Laser-induced fluorescence
Light-controllable dithienylethene chiral switches, 18
Light radiations, 33
Liquid crystal (LC), 1, 34, 63, 103, 146; *see also* All-electrical liquid crystal sensors; Microfluidic LC sensors; Organic field-effect transistors
 applications, 104
 based devices, 104
 -based immunoassay, 152, 154
 capacitance temperature dependence, 111
 electrical equivalent circuit of LC cell, 108
 molecules, 124
 one-to-many stimulation, 124
 optical cell, 147
 orientation in microchannels, 149–150
 phases, 124
 as signal transducer, 146
 temperature sensors, 104
 with tubing cartridges, 154
Liquid crystal display (LCD), 34
 devices, 104
Liquid crystalline microdroplets, 49; *see also* Photochromic chiral LCs
Longitudinal surface plasmon resonance (LSPR), 53; *see also* Photochromic chiral LCs
Low-pass filter (LPF), 119
LPF, *see* Low-pass filter
LSPR, *see* Longitudinal surface plasmon resonance

M

MAA (3-methyladipic acid), 85
Mesogen-functionalized GNRs (M-GNRs), 54
Mesogens, 1
Mesophase, 124
Metal ion sensors, 91–94; *see also* Optical sensors
M-GNRs, *see* Mesogen-functionalized GNRs
Microfluidic devices, 146
Microfluidic LC sensors, 145, 155
 absorbance or fluorescence emission, 146
 advantages of, 151–153
 anchoring of LC, 147

bright LC regions and concentration of anti-IgG, 153
 detection of surface-bound molecules with LC, 147–149
 DMOAP-coated slide, 148
 easy axis of surface, 147
 effective birefringence, 148
 for gas detection, 153–154
 LC-based immunoassay, 152, 154
 LC optical cell, 147
 LC orientation in microchannels, 149–150
 LC with tubing cartridges, 154
 in microchannels, 150–151
 microfluidic devices, 146–147
 multiplex LC immunoassay, 153
 PDCLC, 154
 sensing dots, 151
Microfluidic sensors, 145; *see also* Microfluidic LC sensors

N

Narrow reflection bandwidth, 3; *see also* Broadband reflection
Near-infrared (NIR), 40
Nematic liquid crystal (NLC), 3
Nematic liquid crystal cell (NLC cell), 105; *see also* All-electrical liquid crystal sensors
 capacitance variation vs. external frequency, 110
 equivalent capacitance of, 114
 frequency variation due to temperature, 116
 homeotropic alignment in, 106
 impedance vs. frequency, 109, 113
 multivibrator circuit with, 112
 permittivity data vs. frequency of applied electric field, 107
 planar alignment in, 105
 temperature dependence of permittivity for, 106
Nematic mesophase, 64
Nematic monoacrylate, 4; *see also* Bandwidth tunable CLC
NIR, *see* Near-infrared
NLC, *see* Nematic liquid crystal
NLC cell, *see* Nematic liquid crystal cell
N*LC sensors, 63, 65; *see also* Chiral nematic liquid crystalline sensors
 detection methods in, 68
 as UV sensors, 74

O

OD, *see* Optical density
OFETs, *see* Organic field-effect transistors
 Optical density (OD), 25

Optical sensors, 85; *see also* Cholesteric liquid
    crystalline polymer networks
  alcohol sensors, 90–91
  amine sensors, 86–88
  amino acid sensors, 86
  Ca^{2+} binding to benzoate sites, 94
  carboxylate salt film, 92
  chiral dopants and diacrylate mesogens, 96
  cholesteric polymer containing azobenzene
    chromophores, 99
  components of polymer composite, 86
  humidity sensors, 89–90
  IPN response to RH levels, 91
  mesogen structure, 93
  metal ion sensors, 91–94
  monomer used to make amine sensor, 89
  pH sensors, 85–86
  polymer and Ba^{2+} ion treatment, 93
  polymer film's optical response, 97–98
  strain sensors, 96–97
  temperature sensors, 94–96
  thermal-responsive cholesteric polymer, 94
  UV-Vis transmission spectrum, 87, 88, 90,
    92, 99
  wavelength change in reflection of water-
    saturated films, 97
  working principle of CLC polymer film, 93
Organic field-effect transistors (OFETs), 123,
    125; *see also* LC-*g*-OFET sensory
    devices; LC-*i*-OFET sensory devices;
    LC-*on*-OFET sensory devices
  field-effect mobility, 126
  layers, 125
  performance of, 125
  structure and operating principle, 125

P

P3HT (Poly(3-hexylthiophene)), 126
Patterning methods, 126
PCA, *see* Penthylcyclohexanoic acid
PCE, *see* Power conversion efficiency
PDCLC, *see* Polymer dispersed CLC
PDMS (Polydimethylsiloxane), 146
Penthylcyclohexanoic acid (PCA), 85
Phase locked loop (PLL), 104
Phase-width modulated signal (PWM
    signal), 119
Photochromic chiral LCs, 33, 38, 55; *see also*
    Photochromism; Photoisomerization
  alkene motor, 46
  azobenzene dopants, 39–42
  BPs, 37
  chiral azobenzene, 40, 41, 43, 44, 48
  diarylethene dopants, 42–44
  diarylethenes, 36
  double-twist structure, 47

fabrication of photochromic chiral liquid
    crystals, 38
flower-opening patterns of
    microdroplets, 50
helicity inducers, 39
for infrared sensors, 51
light-driven chiral azobenezene, 51, 52
light radiations, 33
liquid crystalline microdroplets, 49
M-GNRs, 54
microfluidic techniques, 50
NIR-light-directed self-organized BP 3D
    photonic superstructures, 54
phases, 36–38
photothermal effect, 53
polarizing optical microscopy images, 50
reflection colors and reflection spectra, 53
reflection wavelength of photochromic
    chiral LCs, 37
self-organized 3D superstructures, 49
spirooxazine and overcrowded alkenes,
    44–47
spirooxazines, 36
spiropyrans, 36
UCNPs, 52
upconversion processes, 52
Photochromism, 35; *see also* Photochromic
    chiral LCs
  in 1D CLC superstructures, 38–39
  in 3D chiral liquid crystalline
    superstructures, 47–51
  molecules of, 35–36
  photochromic sensor, 34
Photoinitiation, 76
Photoisomerization, 35, *75, see* Photochromic
    chiral LCs
  of chiral diarylethene dopant, 44
Photostationary state (PSS), 41
Photothermal effect, 53
pH sensors, 85–86; *see also* Sensors
Pitch, 64
  of chiral nematic, 83
Planar chirality, 76
PLL, *see* Phase locked loop
PMMA, *see* Polymethylmethacrylate
Polarized optical microscope (POM), 148
Polymer dispersed CLC (PDCLC), 154
Polymer-stabilized CLCs (PSCLCs), 10
  transmission and reflection transforming
    mode of, 16
Polymethylmethacrylate (PMMA), 126, 146
POM, *see* Polarized optical microscope
Power conversion efficiency (PCE), 25
PSCLCs, *see* Polymer-stabilized CLCs
PSS, *see* Photostationary state
PWM signal, *see* Phase-width modulated
    signal

R

RCP, *see* Right-circularly polarized
Reflection bandwidth, 3; *see also* Bandwidth
    tunable CLC
    chemical structures of materials, 11
    cholesteric diacrylate, 4
    cholesteric polysiloxane oligomer, 7
    circularly polarized light transmission, 4
    CLC gel film with broad bandwidth, 8
    film preparation, 11
    fracture surface of pitch-gradient cholesteric
        network, 4
    ND concentration profiles, 11
    nematic monoacrylate, 4
    photoinitiator, 4
    polymer network of polymer-stabilized
        CLC, 6
    preparation of N*LC composite film, 9
    transmittance of polymer-stabilized CLC, 6
    UV-absorbing dye, 4, 5
    UV-light-intensity gradient, 4, 5
    vitrified CLC structure, 7
Relative humidity (RH), 90
Reverse isomerization process, 18
RGB (Red, green, and blue), 54
RH, *see* Relative humidity; Right-handed
Right-circularly polarized (RCP), 18
Right-handed (RH), 1

S

SAMs, *see* Self-assembled monolayers
Scanning electron microscopy (SEM), 10, 126
Selective reflection band (SRB), 84
Self-assembled monolayers (SAMs), 147
Self-organized 3D superstructures, 49
SEM, *see* Scanning electron microscopy
Sensing, 103
Sensors, 65, 103; *see also* All-electrical liquid
        crystal sensors; Chemosensors; Chiral
        nematic liquid crystalline sensors;
        DCL-LC-*g*-OFET sensory devices;
        Microfluidic LC sensors; N*LC
        sensors; Optical sensors; Ultraviolet
        sensors
    advanced, 123
    alcohol sensors, 90–91
    amine sensors, 86–88
    amino acid sensors, 86
    based on photoracemization, 79
    benefit of, 67
    categories, 66
    conventional, 123
    humidity, 72
    metal ion, 91–94
    pH sensors, 85–86
    strain sensors, 96–97
    temperature, 94–96, 104
SmA-like shortrange ordering (SSO), 8
Spirooxazine, 36, 44; *see also* Photochromic
        chiral LCs
SRB, *see* Selective reflection band
SSO, *see* SmA-like shortrange ordering
Sterol-based D provitamins, 76
Stimuli-responsive reflection band, 10; *see also*
        Bandwidth tunable CLC
    chiral switches, 16, 17, 18
    molecular arrangements, 15
    polarizing optical microscope textures, 13
    reflection spectra, 13
    reflection wavelength of light-driven chiral
        molecular switch, 17
    reflectors, 12, 14
    reverse isomerization, 18
    transmission and reflection transforming
        mode of PSCLC, 16
    transmission spectra, 14
    wavelength and bandwidth of selective
        reflection peak, 13
Strain sensors, 96–97; *see also* Optical sensors

T

TAC, *see* Triacetyl cellulose
TADDOL, *see* Tetraaryldioxolanediol
TCOs, *see* Transparent conducting oxides
TEM, *see* Transmission electron microscopy
Temperature–frequency converter, 111; *see also*
        All-electrical liquid crystal sensors
    characterization, 112–116
    design, 111–112
    electronic circuit, 111
    equivalent capacitance of NLC cell 6, 114
    experimental setup for, 114
    multivibrator oscillator circuit, 112
    output frequency as function of temperature,
        115, 116, 117
Temperature-phase converter, 116; *see also*
        All-electrical liquid crystal sensors
    components in phase measurement
        scheme, 120
    electronic scheme of phase measurement
        system, 118
    experimental and theoretical output
        voltage, 119
    maximum sensitivity of phase measurement
        system, 118
Temperature sensors, 94–96, 104; *see also*
        Optical sensors
Tetraaryldioxolanediol (TADDOL), 71
TFA, *see* Trifluoroacetyl
TMA, *see* Trimethylamine
Transmission electron microscopy (TEM), 5, 150

Transparent conducting oxides (TCOs), 125
Transparent polymers, 146
Triacetyl cellulose (TAC), 95
Trifluoroacetyl (TFA), 69
Trimethylamine (TMA), 87

U

UCNPs, *see* Upconversion nanoparticles
Ultraviolet (UV), 3, 35
Ultraviolet sensors, 74; *see also* Chiral nematic
        liquid crystalline sensors
    azobenzene structure, 75
    chiroptical switching, 76
    enantiomeric excess of dopant, 78
    effect of exposure on cholesterol iodide and
        cholesterol nonanoate, 77
    structure of isomers, 76

transformation of provitamin D_2 to
        vitamin D_2, 77
UV sensing, 75–78
Upconversion nanoparticles (UCNPs), 52;
        see also Photochromic chiral LCs
UV, *see* Ultraviolet

V

VOCs, *see* Volatile organic compounds
Volatile organic compounds (VOCs), 66, 153

W

Washout/refill techniques, 21, 22; *see also*
        Broadband reflection
WGM, *see* Whispering-gallery-mode
Whispering-gallery-mode (WGM), 50